TELLING THE BEES

and Other Customs

TELLING the BEES

and Other Customs

THE FOLKLORE OF RURAL CRAFTS

Mark Norman

For Alyssa, who might read it one day

Front cover artwork and internal chapter headings: Tiina Lilja
Back cover photography: Cat Burton

First published 2020
This paperback edition published 2023

The History Press
97 St George's Place, Cheltenham,
Gloucestershire, GL50 3QB
www.thehistorypress.co.uk

British Library Cataloguing in Publication Data.
A catalogue record for this book is available from the British Library.

ISBN 978 1 8039 9261 7

Typesetting and origination by The History Press
Printed and bound in Great Britain by TJ Books Limited, Padstow, Cornwall.

CONTENTS

ABOUT THE AUTHOR

Mark Norman is a folklore author, researcher and the creator and host of *The Folklore Podcast*. Since its launch in July 2016, the internet-based show has grown to sit within the top 10 per cent of its subject area worldwide, having been downloaded over three-quarters of a million times by its listeners.

Mark's aim, with the podcast and his writing, is to try and present a fascinating social history – the beliefs, traditions and customs of our communities and cultures – in an accessible and enjoyable way to a wide audience. He hopes that everyone will come to realise that there are many parts of what we do, think and say that have far more meaning that we might imagine ... until we pause for a moment to consider them.

His wide areas of interest are in these old stories of superstitions and folk practices, but he also holds the UK's largest archive of sightings, traditions and eyewitness accounts relating to apparitions of ghostly black dogs, the folklore of which inspired Sir Arthur Conan Doyle to write arguably his most famous Sherlock Holmes novel.

Mark lives in Mid Devon with his daughter, who is at one with the horses, and his wife Tracey: a historian, author and actor (and awesome manuscript proof reader) who researches witch trials, and as a result knows far too much about the unpleasant things you can do with earthworms. Together, they all share their space with an insane hamster and a feline trip hazard.

INTRODUCTION

Look around you. What do you see? Furniture, perhaps. Clothes, tools or maybe some food and drink. How many of these things were made by hand? Were they crafted by an expert in the field? Or were they mass produced to a formula, in order to satisfy economies of scale: high demand and the need to keep prices low?

Supply and demand have always been with us, but the fast turnaround mass manufacture that we see today has not. Our ancestors worked in a completely different way. They raised animals using traditional methods. Artisan workmen and women would fashion clothes, tools and the like by hand. Their skills were often such that they were revered in the community. Indeed, some thought that their abilities must have come about through a form of supernatural intervention. Maybe from above – or possibly from below.

These were the people involved in rural crafts. Today, there is a distinction between crafts such as these and handicrafts, or crafts undertaken as hobbies or for artistic purposes.

For our forebears, what we now call traditional crafts were simply skills needed to live and survive – to clothe and feed a community. When a set of practices such as those covered in this book – knitting, spinning and weaving, beekeeping, blacksmithing, brewing, milling and baking – are central to a community and viewed with such importance, it is natural that over time a number of beliefs, superstitions and customs will form around them. Looking at them from the outside, this lore can tell us much about the past.

What is folklore? The dictionary would tell you that it is a collection of traditional beliefs, stories or customs, which are passed from person to person in an oral fashion by retelling. The term was first used in 1846, by the English writer William John Thoms, as a replacement for the more cumbersome 'popular antiquities'. Broken down into its simplest form, it is the traditional knowledge (lore) of a group of people (folk). These two components were originally split: folk-lore. Over time, the hyphen was lost and the word became one.

Folklore is a shared culture within a group of people. This is why it still resonates with us so strongly today. The hypothesis that we have some kind of collective folk memory in our subconscious, which draws on the beliefs of the past, certainly goes a long way to explaining why we often seem to just 'know' things. It probably also goes a long way to explain why, outside of academia, folklore has seen something of a resurgence recently.

There are many groups on social media for people to share folklore beliefs and notes, one of the most well known now being Folklore Thursday on Twitter. We still employ many superstitions automatically. Some of us don't walk under ladders. Some of our streets don't have a house number 13, our hotels might go from floor 12 to 14.

At the same time as we find these common themes in our beliefs, we find interesting variations between one place and another geographically. Local variations on a broader theme are common. This happens both because of the natural migration of people, from town to town on the smaller scale and across countries on the larger, and also because these traditional practices and beliefs are so embedded in our surroundings. Folklore gives us a sense of place.

Because stories and beliefs were transmitted through word of mouth, they naturally changed and developed over time. It is kind of the ultimate game of Chinese Whispers. Also, the world around us changed and developed too. Even today, with information being transmitted in different ways, new beliefs emerge and old ones are re-moulded.

Through all of that, folklore – the beliefs of the people – is deep within us, whether we realise it or not. We interact with it every day. This is the reason that my podcast examining these subjects, The Folklore Podcast, (in its fourth year at the time of writing) bears the tag line, 'recalling our forgotten history and recording the new'. The stories of today are shaped by those of yesterday and become the beliefs of tomorrow.

Each of the chapters in this book will give a brief overview of the older mythologies surrounding each of the crafts explored, looking at the patron saints or the gods associated with that profession, before moving on to examine the folk tales, superstitions and beliefs that grew up around them. Some you may know well. Many could surprise you. And just a few might seem a little odd now.

Folklore is such a rich and varied field that it would be impossible to cover every area of the main chapter subjects in one book. The chapter on brewing, for example, examines the process only in relation to beer and ale as a product of the trade. There is equally as much folklore about wines and spirits, but they are not included here. The chapter on milling and baking looks at bread but does not take in cakes. Especially in relation to the fairies, that would be a whole other book! So you will inevitably find that there are beliefs or customs missing that you thought would have been there. You may also have superstitions of your own that you don't think are especially common. I would encourage everyone to try and find a way of recording these as best they can. Cultures are changing and developing all the time, and in the same way that the old beliefs in this book are fascinating to us now, so our beliefs in the modern day will be equally interesting in the future. And, of course, we can constantly build on the archives of beliefs that we already have.

To that end, I am always happy to receive people's memories about the old customs, traditions and superstitions that they remember from their family, which I can add to the archives already existing. Simply look for The Folklore Podcast online and you will find plenty of ways to send them in.

I hope that you find much to interest you, whether you now follow one of these crafts as a profession, or a hobby, or are just fascinated by the way that we lived our lives in the (not so distant) past.

Mark Norman
September 2019

1

WOOL, THREAD AND CLOTH

Swiftly turn the murmuring wheel!
Night has brought the welcome hour,
When the weary fingers feel
Help, as if from faery power;
Dewy night o'ershades the ground;
Turn the swift wheel round and round!

Now, beneath the starry sky,
Couch the widely-scattered sheep;--
Ply the pleasant labour, ply!
For the spindle, while they sleep,
Runs with speed more smooth and fine,
Gathering up a trustier line.

Short-lived likings may be bred
By a glance from fickle eyes;
But true love is like the thread
Which the kindly wool supplies,

When the flocks are all at rest
Sleeping on the mountain's breast

Those words were written by the poet William Wordsworth in his 'Song for the Spinning Wheel', and they represent both a trade and an art that goes way back into the history of peoples the world over. The skills of weaving, spinning and knitting were vital to clothe and keep warm members of every class, race, religion or social group from the poorest to the richest. And so, we find wool, yarn and thread and the working of those materials rooted very deeply in the folklore of countries around the globe.

MYTHOLOGY AND HISTORY

In the Norse myths, the World Tree Yggdrasil stood at the centre of the spiritual cosmos. It connected the Nine Worlds (and the rest of the cosmos) with its roots and branches, and the health of the Universe depended on the health of the tree.

Yggdrasil was watered and cared for by the three most important of the Norns – the female beings who controlled destiny. As they were responsible for the destiny not only of humans, but also of the gods, this made them naturally the most powerful beings in the mythos. Although the Norns also occur in stories drawing wooden lots or casting runes, they are most commonly portrayed seated at the base of Yggdrasil, spinning the threads of fate. The origin of the term 'norn', while not known for certain, is thought by some to derive from the verb 'to twine', which has obvious connotations to the image of the women as spinners of destiny.

What the physical material might be that they are spinning in the artistic renderings of the Norns is debatable. The only mention of the sheep as an animal within the Norse myths is in relation to the watcher god Heimdall, who was said to have hearing so keen that he could hear the wool growing on the sheep's backs. In fact, the sheep is not common as an image in pre-Christian cultures, becoming more prevalent later in connection to terminology such as the 'Lamb of God', for example. The goat was spoken of far more often in the Norse mythos. The goat Heidrun eats the leaves that grow on Yggdrasil and, because of their magical properties, produces mead instead of milk. Thor's chariot is pulled by two goats rather than horses. So perhaps the Norns are spinning goat hair rather than sheep fleece.

In many sources, the species of tree for Yggdrasil is not mentioned, but Old Norse literature often states that it was an ash. In the Norse creation myth, man and woman originate from the trees: Ask from the Ash and Embla from the Elm. In the Christian creation story, the first man and woman are Adam and Eve, which suggests an obvious parallel.

These two characters then appear in one of the oldest English rhymes. In 1381, English Lollard priest John Ball was imprisoned in Maidstone in the county of Kent. Not long after the start of the Peasants' Revolt, he was freed by Kentish rebels, to whom he preached an open-air sermon at Blackheath. His sermon included the passage:

When Adam delved and Eve span,
Who was then the gentleman?
From the beginning all men by nature were created alike, and
our bondage or servitude came in by the unjust oppression of
naughty men.

In this quote, Ball is essentially gendering types of work.
The word delved in this case is referring to the digging of the
earth – in other words, when Adam farmed the land. This is
a male-oriented description of work. So, in referring to Eve
spinning, this is again representative of Ball describing work
from a female perspective. Spinning (more so than weaving)
was generally considered to be women's work. And because
so much thread was needed to make cloth, women spent a lot
of time spinning. It did not, however, hold the same status as
skilled work such as weaving did.

In modern times, gentleman is a term that can be applied
to any male, whereas in Ball's era it was very much a term
used to imply a man of means who did not need to work. The
word gentry comes from the same root. What John Ball was
actually saying in his sermon, therefore, was that in earlier
days when everyone worked to survive, no one person was
better than another and there was no class divide. He was
not suggesting in literal terms that Eve actually sat and span.

There is no evidence in the book of Genesis to suggest that
Eve span. The iconography of Eve with a distaff and spindle is very much the creation of medieval-Christian visions
of the Bible characters. We may not be able to pinpoint its
precise origins, although illuminations as early as psalters
in the thirteenth century depict Eve in this way. Spinning

was often represented as a virtuous activity in illuminated manuscripts.[1] Even the Virgin Mary is shown spinning in two fourteenth-century Italian examples. The same reading may similarly be given to representations of weaving. We can be certain that the skills of spinning and weaving hold such importance that deities who are revered for these skills occur in many early mythologies and beliefs.

The Goddess Weaver in ancient Chinese stories was the daughter of the Jade Emperor and the Celestial Queen Mother. She was said to have woven the stars and their light, which could be seen crossing the sky in the form of the Silver River. Today we call this the Milky Way. The Goddess Weaver was said (in a folk tale sometimes called the Weaver Girl and the Cowherd) to have descended to earth from the Celestial Court some 4,000 years ago. Here, she fell in love with a mortal man: a cowherd.

Niulang, the cowherd, was an orphan who was treated badly by his older brother and his wife. They sent him to live alone, giving him only an oxcart and an old buffalo. The buffalo and its master depended upon each other, but the cowherd was very lonely.

One day, the buffalo miraculously spoke to Niulang, telling him that a band of fairy women from heaven would be found swimming in the lake, and that if he took the red clothes from the bank then their owner would become his wife. He went to the lake and did so. Seeing him, all of the celestial women quickly dressed and flew to heaven except for Zhinu, the one whose clothes Niulang held. He asked her to marry him and she quickly agreed, and the couple fell very much in love.

The cowherd and the Goddess weaver lived happily together. He tilled the land and she wove cloth. We may notice here the similarity to the thirteenth-century psalters and John Ball's poem regarding Adam and Eve. This suggests that the gendering of these tasks is universal. One day, the buffalo warned his master that he was about to die, and instructed him to keep his hide for a later emergency. Niulang was very sad but did as he was instructed.

Shortly after this event, Zhinu's grandmother in the heavens learned of the marriage to a mortal and was very angry. She was the queen of the emperor of the heavens, and as such sent down gods and soldiers who quickly reclaimed Zhinu. Seeing this happen as he returned from the fields, Niulang put their children into baskets, put on the buffalo hide and flew to the heavens in pursuit.

Here the versions of the tale vary. In one version, the grandmother separates the lovers and Zhinu threatens to stop weaving the Silver River, which would threaten heaven and earth with darkness. In the other version, the Silver River is used to keep the lovers apart.

Either way, as a concession, the cowherd and his children live on one side of the Silver River and his wife on the other. They are allowed to reunite once a year, on the seventh day of the seventh moon. Since ancient times, as this tale is one of the great Chinese love stories, this date has been the effective Chinese Valentine's day.

Rather than being on a fixed date as it is in Western cultures, being related to the lunar cycle in this way means that the Chinese celebration has a variable date. Known formally in China as Qi-Xi (literally translating as seven night) the event sometimes goes by the name of the Double Seven Festival.

The weaver in this tale is associated with the star Vega (known as 'Weaving Maid') and the cowherd with Altair. These two stars sit on either side of the Milky Way and are easy to spot in the summer sky, being the fifth and eleventh brightest stars respectively. Altair is adjacent to two small bright stars, Alshain and Tarazed, which represent the children of the couple.

It is said that at this time you will not see magpies on earth, because they fly to the heavens to build a bridge between the lovers. By this method the weaver and the cowherd are able to meet. In some provinces of China, during the Dragon Boat Festival that is celebrated earlier in the year, five-colour silken ropes are woven. At Qi-Xi, girls throw these ropes onto the roofs of houses for the magpies to collect. The birds are then said to carry these ropes off to build the bridge.

As you would expect, there are a number of traditions associated with the festival of Qi-Xi. Some Chinese people believe that adorning an ox's horn with flowers, in honour of the ox in the legend, will prevent bad luck. It is traditional on the night of Qi-Xi for women to wash their hair. The following morning, children use water left overnight in the back yard to ensure a beautiful appearance.

Chinese Valentine's Day is also sometimes called 'The Daughter's Festival', because it is a day for unmarried girls to seek love. Traditionally, Chinese girls desired to develop good needlecraft skills like the Goddess Weaver. These skills were considered to be vitally important for family life, and so on Valentine's Night in China, unmarried girls would pray to the weaver's star, Vega, to be granted good handicrafts. When the star was high in the sky, the girls would place a needle on the surface of some water. If the needle floated then the girl's skills were thought to be sufficient and she was ready to marry. Contests would also be held where girls would try to be the best at threading needles in low lights.

In the Tang Dynasty in China, the weaver goddess was said to have come to earth with her two attendants, whereupon she showed a court official that the robe of a goddess has no seams, because it is created on a loom rather than by needle and thread. The phrase 'a goddess's robe is seamless' became embedded in the language of the culture, representative of an example of perfect workmanship.

We may also find a weaving theme in a tragic Japanese folk tale, *Tsuru Nyobo*, translating as 'Crane Wife'. Again there is more than one version of this.

One day, while working on his farm, a white crane falls from the sky in the fields where a young man is working. He notices that it has been hit by an arrow. He cares for the bird and cleans its wound before returning it to the skies. Later that day, when he arrives home, he finds a beautiful girl at his hut, whom he has never seen before. She says that she is his wife and, despite protestations that he cannot support her, she insists that she has plenty of rice and prepares dinner. They continue to live together and the sack is always full of rice.

One day, the woman asks her husband to build her a weaving room, but instructs him that he must never look inside. She shuts herself away and emerges seven days later, very thin but with a beautiful piece of cloth, which she says will fetch a good price at market. The man sells it and indeed it does make a lot of money.

The woman again shuts herself away to weave, but the man cannot contain his curiosity as to how she weaves with no thread, and he looks inside the room. There he sees that the woman has gone and in her place is the crane, pulling out its feathers and using them as thread to weave. Having been seen in her true, pitiable state, the crane flies off leaving only the cloth to remember her by.

A second version of this tale is known as *Tsuru no Ongaeshi* or 'The Crane's Return of a Favour'. Here the role of the man is replaced by a poor elderly couple but the rest of the story remains very similar with the elderly man releasing the crane from a trap in which it has been caught. It is essentially a variant of the 'curiosity killed the cat' moral.

WEAVING

The character of a weaver is often found in folklore as a ghost within a community who needs to be removed. The reason for this is that, alongside other such skilled craftsmen as tailors or millers (as we shall see in a later chapter) the ability of the weaver made them stand out in a community. As with many folk tales of ghosts, characters such as these were seen to wield some power over their community. Sometimes this might be because they are rich due to their trade, or because of rumours that they made a deal with the Devil in order to obtain their skills. Inevitably, the people around them become mistrustful of them and hostile towards them because of their actions. However, because they are too formidable a character to tackle head-on in life, it is after they die that the community seeks vengeance against them.

By way of illustration, we might take the story of a weaver named Knowles, who lived in the hamlet of Dean Combe, situated in an area of the South West of the United Kingdom known for its cloth manufacture. Knowles was said to be one of the most skilled weavers in the area, consistently producing the finest knap. Because of his skills, Knowles became a rich man, but he was not very generous with his wealth. He was also hated by his neighbours, who said that he was an evil, selfish gossip.

Knowles worked from sunrise to sunset each day, eventually dying whilst sat at his loom in the loft. Despite being ill-liked in the parish, many local residents came to the man's funeral. This was more to wish well to the weaver's son Fernley than it was to honour the old man. Fernley was a much more pleasant and respected man in the parish and he knew that he would now have to prove that he had the skills that his father had taught him at the loom.

Some time passed and the funeral came and went. Fernley knew that he must start working, and so descended to the kitchen to light the fires and take some breakfast before he began. Whilst sitting at the table, Fernley was shocked to hear the familiar sound of the loom thumping upstairs in the loft. He crept up the stairs to investigate, thinking that perhaps someone had broken into the house. Easing the door slightly ajar and peering in, however, a rather different sight met his eyes. There at the loom sat the ghost of his father, the old weaver, working as ever he had done.

Egyptian mat loom in use. Source: 'Yarn and Cloth Making: an economic study',
Kissell, M.L. 1918. (Public domain)

Alarmed and confused, Fernley could think of nothing to
do but fetch the local vicar, who came to the house thinking that he would need to employ the traditional bell, book
and candle. Arriving at the house, the vicar remained on the
ground floor and shouted up to the weaver to leave the property and return to his proper place in his grave.

'I will as soon as I have finished this quill,' cried out the
spirit of Knowles in an unearthly voice.

The priest again commanded the spirit to leave the house
and return to the graveyard. This time the spirit came

downstairs, at which point the priest threw holy water into its face whilst reciting a prayer. With a scream, the image of the weaver transformed into a ghostly black dog. Through use of the Bible the priest brought the dog to heel and commanded it to follow him, where he led it down the lane to the local woods. There the priest banished the spirit to a task. Picking up an acorn shell by an oak tree, the priest told the dog-form of Knowles that he would find eternal rest only when he had emptied the pool in the woods using the acorn shell.

Legend says that the local people will not visit the pool at either noon or midnight because they will see the image of the terrible black dog still trying desperately to empty the pool with the acorn shell so that it can find its rest.

SPINNING

Weaving, if you are not a bird like the Crane Wife was, usually begins with spinning. We have already seen that in the late Middle Ages, it was usually thought that spinning was a female role, whereas weaving was more of a male tradition. It is actually the case, however, that men have taken on what was once a more inherently female calling. Certainly, among the pantheons of the gods, only the goddess characters wove. In modern societies across the mid-parts of Asia, weaving is associated with women. In more ancient history there are more definite gender distinctions, and Herodotus noted that among the Egyptians, for example, it was the men who wove:

The Egyptians in agreement with their climate, which is unlike any other, and with the river, which shows a nature different from all other rivers, established for themselves manners and customs in a way opposite to other men in almost all matters: for among them the women frequent the market and carry on trade, while the men remain at home and weave; and whereas others weave pushing the woof upwards, the Egyptians push it downwards …[2]

He does note, however, that the gendering of the tasks in this culture is the exception rather than the rule.

The spinning wheel as a tool to aid in the production of yarn has an uncertain history. It has been suggested that its origins come from India at any point between AD 500 and 1000,[3] and that it subsequently spread throughout China and into Europe, where it started to become recognised around the fourteenth century. It was not popularised in Western cultures until much later. Prior to this, all spinning would have been undertaken using a distaff and spindle. We find the distaff as an emblem in various pieces of folklore. In English, the term 'distaff side' indicates relatives on the maternal side, linking in with the idea of spinning being a female pursuit. The male side of the family is known as the spear side.

The seventh day of January, the first free day after the twelve days of Christmas, is often celebrated as 'St Distaff's Day'. We find it noted in *Chambers Book of Days* that on this day the women would return to spinning, having either nothing else to do with their time after the rest period, or undertaking it between bouts of other more serious work.[4]

They probably didn't work too hard at it on this day, however – the ploughboys were not hugely motivated to go back to work and took it upon themselves instead to set fire to the flax that was produced. In response to this prank the maids would douse the men with water from their pails. We might assume that the Christmas spirits had not quite diminished by the thirteenth day!

The poet Robert Herrick recorded St Distaff's Day in a short stanza:

> Partly work and partly play
> You must on St Distaffs Day:
> From the plough soon free your team;
> Then cane home and fother them:
> If the maids a-spinning go,
> Burn the flax and fire the tow.
> Bring in pails of water then,
> Let the maids bewash the men.
> Give St Distaff' all the right:
> Then bid Christmas sport good night,
> And next morrow every one
> To his own vocation.

Humour aside, spinning was a very important skill and the act of spinning and the distaff itself became synonymous with the womenfolk. An unmarried woman was referred to as a spinster, for example. In France, the proverb came about that, 'the crown of France never falls to the distaff'. Whilst a woman would not be able to make a full living from spinning, according to Anthony Fitzherbert in his *Book of*

Husbandry, it 'stoppeth a gap'.[5] In fact, no rank was too high to use a spindle and distaff.

In the book of Proverbs, King Solomon spoke of women laying their hands to the distaff (Proverbs 31 verse 19), and its use can be traced on the monuments of ancient Egypt. In pre-dynastic Egypt, the goddess Neith, as well as being a war goddess was also one of weaving. She is known by other similar names, including Nit. Whilst I would love to be able to tell you that the etymology of knitting comes from this root, sadly it is not the case – we will move to knitting in a while. In his book *The Gods of the Egyptians*, however, author E.A. Wallis Budge does note that the root of the word weaving (ntt) is the same as for the word 'being'.[6] Nit is portrayed in Egyptian iconography by various emblems, one of which is the weaving shuttle.

Early seventeenth-century woodcut of a woman spinning. (Creative commons licence)

One of the oldest of the Greek mythological tropes features the Moirai, or Fates, as three crones who spin the thread of human destiny. Whilst this is a more allegorical representation of thread on the distaff, we find weaving and spun thread cropping up elsewhere in the mythological tales. In Crete, Ariadne had spun the thread that Theseus used to safely navigate the maze of the Minotaur. In the Olympian tales, Athena is the weaver goddess who lost to Arachne in a weaving competition. The latter's punishment was to be turned into a spider, so that she could spend the rest of her life weaving. There are many more examples throughout Homer's Odyssey and into Roman literature.

The figures of the three fate deities occur in similar forms but by different names in the mythologies of many cultures, with frequent references to the acts of spinning. We began this chapter by mentioning the Norns, who span at the roots of the World Tree, Yggdrasil. In Romanian folklore, where spinning is also associated with fate deities, there are many taboos that work to restrict the activity at different times. Sometimes the taboos pertain to a household situation – for example, it is forbidden in Romanian lore to allow wool to be spun in a house where a woman is in labour. To do so will supposedly bring bad luck and the baby could be suffocated by the umbilical cord when it is born.[7]

These beliefs are also associated with particular days connected with the deities. The Romanian spinning deity Joimarita (whose name comes from the Romanian word for Thursday, joi) punishes women who have been lazy and not finished their spinning on time. The deity associated with

Tuesday in Romania, Martolea, is connected rather with the fact that Tuesday is a semi-holy day. It therefore forbids four domestic activities from being undertaken on this day, one of which is the spinning of wool. Martolea is also known as the Demon of Tuesday or sometimes Mart Sara, which literally translates as Tuesday evening, the time at which it comes down from its mountain home.

Mart Sara is a little like a siren in that it lures people with its singing. Being able to change shape, it is able to appear to potential victims in a form that they will find particularly appealing: unmarried women will see it as a young man and married men as a virgin. As with much of the folklore that exists to prevent certain activities from taking place at significant religious times, the punishments meted out by Mart Sara can be rather severe. Unmarried women who do not observe the semi-holy day are killed by having their stomachs ripped open, their guts hung on nails or scattered around the dishes. Married women may find either their husband or their baby killed.

In some areas, rewards are given for observance of the day. In Bucovina, these take the form of eggs or flowers left on the doorstep. Women who wear the spring token, known as Martisor, on the first day of March will receive a silver coin from the deity. They should keep this coin for the next year.

A Martisor token takes the form of red and white cord twined together. The red signifies the winter, and white the spring. Other symbols of good luck can be attached to the white cord, some of which we may recognise as similarly lucky in other cultures, such as the clover or the chimney sweep. This spring practice seems to be very old – 8,000 years from suggestions found in archaeological excavations. Certainly in Roman times, Martisor took the form of white and red pebbles on a string.[8]

The Martisor custom may also be found associated with the legend of Baba Dochia, in which spinning also plays a part. Again, Dochia's symbolic death and rebirth ties in with the common ritualistic spring traditions. In some versions, Dochia is said to have spun threads whilst driving sheep into the mountains.

Superstitions are found, of course, with all crafts and we may often observe the links to religious days or times. In Denmark, Sunday is the day when spinning was said to be forbidden. Similarly, if you spun on Good Friday then you risked inflaming your fingers. Lent and Easter are common times when items or practices are given up temporarily. It was also said that at Christmas time, nothing was allowed to turn – particularly not the spinning wheel.[9]

The mythologies of sun goddesses are found in many areas, and usually resemble Sol in many aspects, albeit with their own cultural signifiers. The imagery of spinning is the core, with the sun representing the spindle and its rays the thread. It has been suggested that this is the reason why it was not permissible to turn or dance anticlockwise, known traditionally as widdershins. Doing so would untwist the yarn.[10]

In the Baltic countries, the sun goddess is Saule – a name not too dissimilar to Sol. She is signified by a wheel and spins the sunbeams. Baltic legends have taken on various aspects from other regions, including Greek, and we find much legend to explain astronomical aspects of the heavens.

Saule, my amber weeping Goddess
creating light like thread
As 'Saules Mat' my mother sun, daily blessing
your thankful world with light

This verse is found in the traditional folk songs, or *dainas*, of the Baltic states. Amber is also known as the sun stone. Crying tears made of the gemstone is symbolic, demonstrating that even the tears of a deity have virtues that cannot be found in mortals. Ancient burial mounds have been found containing spindles made from amber. Although in many cases these appear to be votive, some of the ones in Baltic areas seem to show signs of having been used for their intended purpose.

The Lapp people of the Scandinavian Arctic regions had a similar sun goddess named Beiwe, taking as its root the regional name for the sun. It was known at the times of major festivals for flax and a spinning wheel to be placed on an altar as offerings to Beiwe.[11]

This goddess appears as Peiwe in the Kalevala – the Finnish epic poem of the nineteenth century, compiled from the oral folklore and mythology of Karelia and Finland. It contains a number of references to spinning and weaving goddesses. For example, Rune VIII – Maiden of the Rainbow – begins:

Pohyola's fair and winsome daughter,
Glory of the land and water,
Sat upon the bow of heaven,
On its highest arch resplendent,

In a gown of richest fabric,
In a gold and silver air-gown,
Weaving webs of golden texture,
Interlacing threads of silver;
Weaving with a golden shuttle,
With a weaving-comb of silver;
Merrily flies the golden shuttle,
From the maiden's nimble fingers,
Briskly swings the lathe in weaving,
Swiftly flies the comb of silver,
From the sky-born maiden's fingers,
Weaving webs of wondrous beauty.

In later European folklore, both weaving and spinning keep connections with magical properties. The character of Mother Goose, a traditional teller of stories, is sometimes also associated with weaving – undoubtedly because of the connection of spinning yarns that go with storytelling. Mother Goose is an interesting character, whose origins are somewhat shrouded in folklore. The title is one given to an imaginary teller of fairy stories – usually the archetype of a rural woman. No individual author is associated with the character and the first appearance cannot be put to one time or person. She is sometimes known as Bertha the Spinner, or the Goose-Footed Queen. This title is said by some to have been given after the wife of King Robert II of France, who was a great storyteller – although the great folklorist and authority on Mother Goose Iona Opie did not hold with this.

The same title, Queen Goosefoot, has also been ascribed to Bertrada II of Laon, the mother of the first emperor of the

Holy Roman Empire, Charlemagne, before King Robert in the eighth century. It was not used for Bertrada at the time, but was rather coined by the French minstrel Adenes Le Roi in the thirteenth century. Additionally, suggestions that the name refers to Bertrada being born with a deformed foot may also not be accurate; Le Roi does not mention this in his poem. An alternative theory is that the name refers to the German goddess Perchta, to whom we will return shortly.

An apocryphal theory from America holds that Mother Goose was a real person. Elizabeth Foster Goose was the second wife of Isaac Goose. The legend tells that Mrs Goose used a collection of songs and rhymes, which her son-in-law subsequently published in 1719, to entertain her grandchildren. No evidence has ever been found for this publication and there are certainly no existing copies, which naturally casts doubt on the tale. A grave in the Old Granary Burying Ground in Boston, Massachusetts, bears the name of an earlier Mary Goose. Once again, there is nothing to link her to the character of Mother Goose aside from the unusual surname. The vague legend continues to be perpetuated, as people visiting the grave leave coins as an offering.

In modern Western culture, Mother Goose is most commonly associated with the pantomime of the same name. The story comes from an ancient Greek legend about a goose that lays golden eggs – one of Aesop's famous fables. This makes it probably the oldest tale to have been adopted and used in a pantomime format.

The motifs of spinning are very strong in a number of other fairy tales. We naturally think of Sleeping Beauty here. In all the various forms of this story (as fairy tales always exist in many variants), the sleeping curse is activated by the pricking of her finger on a spindle. Likewise, the tale of Rumpelstiltskin, wherein the eponymous character aids a girl in spinning straw into gold, is also common and was already old when it was collected by the Brothers Grimm.

In the tale of the Three Spinners, the fourteenth of the stories collected by the Grimms, we find parallels with Rumpelstiltskin. Here, the unwilling girl who should be spinning is aided by three deformed characters who tell how they got their deformities from years of their craft, leading to the girl never having to spin again. We might draw a link between these three and the mythological fates.

The Grimm brothers, German academics of the nineteenth century, were among the most prolific collectors of folklore and folk tales of their time.

Germanic folklore is quite rich with references to spinning and weaving. The Song of the Spear, quoted in the Scandinavian Njal's Saga, is the Battle Song of the Valkyries and was overheard by Daurrud on the morning of the Battle of Clontarf. Daurrud saw twelve folk riding together to a bower and followed them when they went from his sight. Looking through a window slit at the bower he saw that the twelve women had set up a horrific loom on which they were working. The weights were men's heads and the weft was their entrails. They used a sword for a shuttle and arrows for reels, and sang their songs as they worked, one of which started:

> See! warp is stretched
> For warriors' fall,
> Lo! weft in loom

'Tis wet with blood;
Now fight foreboding,
'Neath friends' swift fingers,
Our grey woof waxeth
With war's alarms,
Our warp bloodred,
Our weft corseblue.

In the mythology of Germany itself, there were two goddesses who were associated with spinning and weaving. One was Holda, or Frau Holle, and the other was Perchta, also sometimes known as Frau Berchta or Bertha – note here a link with the name Bertha to the character of Mother Goose. These characters have an association with the Perchten, the Alpine spirits from which the more popular figure of the Krampus was to emerge. Frau Perchta and Frau Holle are essentially the same folkloric character, the latter being found in the geographic area north of Bavaria.

We find the Perchten figuring particularly in celebrations around the time of Epiphany. This takes place on 6 January at the end of the Christmas period, the day prior to the events on St Distaff's Day in other places, as we have already discussed. We, again, have the Grimm brothers to thank for bringing to the fore the folklore surrounding the figures of Perchta and Holle. It was Jacob Grimm, writing in *Deutsche Mythologie*, who drew the connection between the two. Al Ridenour, expert on Alpine lore, notes that many see the two figures as a single entity (Holde-Perchta) and that, furthermore, Grimm also highlights similar figures with combined names, including Berchthold and Hildaberta.[12]

Frau Perchta's house visits were primarily to ensure that domestic spinning duties had been undertaken. It was customary for all unspun flax to be completed in time for Epiphany. Any that remained unworked would be destroyed by Perchta, which is partly what accounts for her also being called 'Spinnstubenfrau', translating as the 'Spinning Room Lady'. Flax had to be completed by this time because the end of the Christmas celebrations were when the construction of the upright loom took place, heralding the start of weaving time.

There is a more general connection between Perchta and domestic untidiness. This is most likely an aspect that was added to the character as time went on, and spinning and weaving moved out of the home into more industrial settings. The watchful eye of Frau Perchta would have been too well established to just fade away, and so the character was appropriated to continue to tie her to the domestic setting. There are a number of punishments associated with Frau Perchta, which were meted out on those who failed in their domestic duties or otherwise offended her. A particularly unpleasant one was to cut open the stomach of the responsible person, remove the contents and stuff the space remaining with waste from the flax and other detritus. An iron needle would then be used to sew up the victim. Another was to stamp on the person offending her. This has led to descriptions of Frau Perchta with oversized feet – an aspect that has also been applied to Bertha and Mother Goose.

The significance of the large or malformed foot in these characters can perhaps be explained in a couple of ways. Grimm thought that in the case of Perchta it may be connected with the spinning aspect, and signify the foot used to power the spinning wheel. There are also myths in many places that suggest having a clubfoot gives a person some sort of link between this world and the spirit world, which may feed into the overall archetype. As an interesting aside, pregnancy lore in Ireland warns any woman who is carrying a child not to walk over a grave. If she does so, then the child will be born with club feet. The Irish name for clubfoot, in fact, is 'crooked churchyard'.[13]

We can find other parallels between the stories collected by the Grimm brothers. For example, in the story of Spindle, Shuttle and Needle, the female protagonist is spinning while chanting a verse to bring a true love to her. In the English, this translates as 'Spindle, my spindle, haste, haste thee away, and here to my house bring the wooer, I pray.'

The spindle, which is magical, flies from the girl's hand and unravels a thread behind it, which a prince follows in order to find the girl that he is destined to marry. A parallel here, of course, with the story of Theseus and the Minotaur in the form of following a thread to achieve a goal.

The Grimms collected folk beliefs and superstitions as well as these tales. In the 1835 book *Deutsche Mythologie*, Jacob Grimm records that:

> If, while riding a horse overland, a man should come upon a woman spinning, then that is a very bad sign; he should turn around and take another way.

KNITTING

From spinning and weaving, we move on to work the thread in a different way with the practice of knitting. As I said earlier, sadly the etymology of knitting does not come from the Egyptian goddess Nit, but rather from the Old English *cnyttan*, which is linked with the Old Norse term *knytja* and Germanic *knütten* amongst others. All of these terms mean to tie, knot or bind together.

Within traditional witchcraft lore and practices, knots are frequently used for the purposes of creating binding spells, and knitting may work along these lines, with glass needles being especially effective. The knot is believed to work as a container for the magic. Additionally, due to the repetitive nature of the act of knitting, it serves as a good method of reinforcing a spell, reciting the intention over and over with each stitch as a way of strengthening the work being done.

Knot magic was traditionally employed by sailors who needed to raise a wind to sail. Generally, a piece of rope or cord would contain three knots. Untying the first would release a gentle wind, the second a strong wind and the third a hurricane. The sailors themselves did not profess to have the ability to place the required magic into the knots tied in the rope. They would have to procure this magically imbued cord from a practitioner who 'sold the wind'. Writing about the subject in his classic 1922 work *The Golden Bough*, Sir James George Frazer offers examples from wizards in Finland and Lapland, and witches in Shetland, Lewis and the Isle of Man.[14] As well as all of these northerly locations, witches on the South-western peninsula of the United Kingdom were also commonly said to do this.

Wool would sometimes be used by witches to assist in binding and trapping something to stop it from being passed on – such as an ailment, for example. The Museum of Witchcraft and Magic in Cornwall holds amongst its collections a wooden example of a 'get lost' box that was constructed for this purpose. To assist in containing the contents, the box was wound round many times with red wool. In a similar way, wishes could be captured in a wish box.

Red woollen thread was a common material for banishment in magical practices, and was used in this fashion in other cultural belief systems, such as in Bulgaria where materials were important in many ritualistic ways. Here, both wool and cotton were considered lucky and, for that reason, were connected to rituals that transition to something new.

One example of the importance of wool in this way is at the time of marriage. In Bulgarian folklore, wool promotes fertility. This power is said to stem from its connection to the earth and therefore, by extension, to the world below. In the World Tree myth, wool is found within the roots. Wedding customs in this country include a ritual circle of individuals who take away the bride. One of these is a woman who spins constantly until the bride reaches her new home. In other parts of Bulgaria, the bride may be given a length of wool, which should be raised three times at the door of the couple's home, or the couple should step on white wool before the marriage ceremony takes place.

Wool could also be used as part of the contents of a witch bottle. An example of one of these, also held in the Museum of Witchcraft, was accompanied by a note from the donor that stated:

> Witch bottles are an old part of the folk magic tradition. They are traditionally used to keep bad spirits or influences away from your house. Some are filled with sharp metal objects and vinegar or urine. Another traditional form of witch bottle is packed with short pieces of thread. The idea is that the threads form a dense maze, which will confuse the spirits so they won't be able to get past the bottle. I have adapted the tradition a little. I do a lot of sewing, embroidery and knitting; mostly I make magickal artefacts. Every time I make something, I snip off the odd ends of thread or wool and put them into a bottle. Thus these bottles, in addition to providing protection, should exude some of the creative and magickal energy that went into their making.

Clothing has always been seen as a status symbol, and knitted garments are no exception. Whilst it was once the case that knitted garments may have been the most expensive possession that someone owned and would have been continually patched or mended, this practice tended to die out. In the same way that stories and meanings within folklore change over time, attitudes to knitted items did too, and darning and mending practices fell out of favour because rather than the garments being seen as lavish, they became a mark of poverty. The act of spinning or knitting, which once spanned classes, became more associated with the poorer and more impoverished end of the social scale. If you consider nineteenth-century fiction for example, you will see these attitudes represented quite clearly. The ladies in Jane Austen's novels would embroider, but not knit. Plain work such as that was the business of the lower class by this time.

The history of knitting, in fact, developed in a similar way to the spread of folklore. The practice of the art itself can be traced across Europe, for example, in the same way as the spread of certain stories and beliefs can. The development of the techniques of knitting grew in ways that were symbolic of the community in which the practitioners were living. Types of stitch were developed to represent those shapes and patterns that were observed in the natural world around the knitter; stitches such as 'Little Leaf Lace', 'Travelling Vine' or 'Tree of Life'. This latter name also carries across to the name of a pattern style, *Yggdrasil*, which, as we saw at the beginning of this chapter, was the name of the World Tree from Norse mythology. The pattern is found in Scandinavian designs, often as a sock pattern, and is derived from the symbol of the mythical tree on the Överhogdal Tapestries. These amazingly well-preserved cloths date from the late Viking period, and were discovered accidentally in 1909 in Överhogdal Church in Härjedalen, Sweden, by a 14-year-old boy clearing out a chest used for firewood.

Early knitting patterns for garments were not written down, but were passed orally from adult to child or from master to apprentice in exactly the same way that early stories or beliefs were handed down and remembered. Some patterns in coastal areas of the United Kingdom were passed on through the migration of the 'herring girls' during the fishing season. These were ladies who travelled great distances around the coast, following the fishing fleet for work.

Amber spindle whorl. Source: The Swedish History Museum, Stockholm (Creative commons licence)

In the same way that stories and folklore eventually began to be recorded in the printed word via pamphlets or chapbooks, so printed instructions for garments also began to become more common in the nineteenth century. This is not to say, however, that there are not written examples before the time. The oldest knitting pattern that we know to have been laid down in written form can be found in a 1655 medical book called *Nature Exenterata: or Nature Unbowelled by the Most Exquisite Anatomizers of Her.*

Traditional styles of knitting are less likely to have been lost in very rural or isolated communities, simply because there would have been less influence from outsiders. It was a craft that would have been the main source of income for many. For this reason, it was common for people to knit whilst undertaking other tasks. Aids were created to assist with this, such as a wooden sheath that was wound around the waist and which supported the wool. One Yorkshire woman, named Slinger, would walk to the local market around 3 miles away, carrying all of the garments her family had knitted to sell that week in a bag on her head. She would continue her own knitting throughout the walk.[15]

Ever since the Industrial Revolution led to cheaper mass-produced garments, the demand for hand-knitted items has remained strong. Part of their appeal, quality aside, is their more individualistic nature. Knitters are more able to respond to trends or changes in taste, and would often create designs that were informed by cultural legends or stories. There was an intrinsic link between knitting and storytelling, when people came together in groups to do so, as evidenced in 1837 when William Howitt wrote:

At Garsdale, the old men sit in companies round the fire, and because they get so intent on knitting and telling stories, they pin cloths on their shins to prevent themselves from getting burnt.[16]

YARN BOMBING

To finish off the examination of this topic, we come up to date and turn to the developing field of urban folklore. As the boundaries between rural life and city life become increasingly blurred in the modern world, we find more and more practices and beliefs that were certainly at one time seen as the domain of the 'country dweller' being absorbed and acted out in the urban landscape. Such things as wassails no longer take place just in orchards, but may be found in urban parks, outside off-licences or along the High Street – as was the case in Tarring, Sussex, in 2019.

Wool has also made its way into the urban sprawl in the twenty-first century, in the form of what we now some-times term 'Yarn Bombing'[17] – that is, the act of covering or decorating parts of the landscape around us with wool-len artefacts. In an article in the *Daily Telegraph* newspaper in January of 2009, guerrilla knitting, as it was originally termed, was noted as being initially almost exclusively about reclaiming and personalising sterile and cold public places. The exact origins of the idea are not 100 per cent certain, as is often the case even with newer folklore. There are certainly examples recorded as early as May 2004 in Den Helder in the Netherlands, and in 2005 knitters in Texas in the United States utilised their left over or unfinished knitting projects for the purpose.

The start of the movement is often attributed to Houston resident Magda Sayeg. She says that she first had the idea in 2005, when she covered the door handle of her boutique with a custom-made woollen cosy. However, earlier than this in 2002, artist Sharon Schollian was knitting stump cosies for felled trees in Oregon. The movement progressed over time, and innovated with the creation of the 'stitched story'. This concept is generally attributed to Lauren O'Farrell from London, who founded a graffiti knitting collective called Knit the City. Their first installation in August 2009 was titled 'Web of Woe'. Lauren did not like the more violent connotations of the term 'yarn bombing' from the American examples, and instead employed the slightly tamer term of yarnstorming to describe the group's activities. The Knit the City collective maintained a sense of humour about the group's origins, and so whenever they were interviewed, the six original members would tell a different story as to how the group came about. To further add to the air of mystery, members used superhero-style street names; O'Farrell's was 'Deadly Knitshade' , while others included 'Knitting Ninja' and 'The Purple Purl'.

In late August of 2009, Knit the City became the first collective to publicise a live yarnstorm on Twitter, involving the six churches of the Oranges and Lemons nursery rhyme. Called 'Oranges and Lemons Odyssey', images of the six-hour installation were published on their Twitter feed in real time.

From small beginnings with a cosy for a wooden barrier in Covent Garden and later what was probably their most well-known piece in the form of a phonebox cosy in Parliament Square, the group have since shown work at the Tate Britain, and had commissions from large knitwear companies and other groups such as the Nintendo Corporation. Knit the City would add paper or fabric tags onto their pieces of work, carrying a logo and their website address, along with the phrase 'Confess Your Theft'. Members of the public were encouraged to take the items away and report back.

We may take much from this example to look at the dissemination of similar pieces of folklore in the landscape today. Yarnstorming has become quite common. Recently we have begun to see a proliferation of painted stones in the landscape, again often bearing Twitter hashtags so that finders can report back. This idea stems from the hobby of Geocaching, amongst other things. And all of this comes from ideas such as the anonymous leaving of offerings on wayside graves or other markers, from love locks and other contemporary assemblages. It all goes a long way to demonstrate how the beliefs and ideas of our ancestors continue to proliferate through our modern lives in many and various ways, and which demonstrate why, rather than being a subject to be sidelined or demeaned, folklore is a subject that should hold an ongoing fascination for us all.

Chapter One Sources

1. Holloway, J.B. et al (Ed), Equally in God's Image: Women in the Middle Ages, 1991
2. Herodotus, The Histories, Book 2, Chapter 35
3. Smith, C.W. & Cothren, J.T. Cotton: Origin, History, Technology and Production, 1999, John Wiley & Sons, New York
4. Chambers, R., Chambers Book of Days, 1864
5. Fitzherbert, J., The Book of Husbandry, 1523. Available online at http://www.gutenberg.org/ebooks/517457
6. Wallis Budge, E.A., Gods of the Egyptians, 1904.
7. Adina Ciubotariu, personal correspondence, 24 April 2018
8. beijing.mae.ro/en/romania-news/375 Accessed 2 May 2018
9. Susanne Lund Peterson, personal correspondence, 26 April 2018
10. Monaghan, P., O Mother Sun! A new view of Cosmic Feminine, 1994, Crossing Press, US
11. Holmberg, U., Finno-Ugric and Siberian, Mythology of All Races IV, 1927
12. Ridenour, A., The Krampus and the Old Dark Christmas, 2016, Feral House
13. Daniels, C.L., and Stevans, C.M. (Eds), Encyclopaedia of Superstitions, Folklore and the Occult Sciences of the World, 2003, University Press of the Pacific
14. Frazer, J.G., The Golden Bough: A Study in Comparative Religion, 1890. Available online at http://www.gutenberg.org/ebooks/3623
15. Walker, G., The Costume of Yorkshire, 1814, Robinson and Sons, Leeds
16. Howitt, W., The Rural Life of England, 1838.
17. www.urbandictionary.com Accessed 17 May 2018

2

BEES AND BEEKEEPING

When the last of the sunlight goes,
and shadows stretching from the shade
of trees and bushes, long hedgerows,
join up together to invade
wild grasses and the flat pasture,
turning from shadow into night,
then the bees, scattered far and near,
take notice, and start on their flight
back to those walls and roofs they know,
beehives where their small bodies rest
between dark and dawn; they go
over the threshold, noisy, fast,
massing in hundreds at the doors,
and pour past into their close cells,
cramming chambers and corridors
while the last of the daylight fails:
sleep silences the working hive
and leaves it quiet as the grave.[1]

The opening lines of this chapter come from Book IV of the *Georgics*, a poem by the Latin poet Virgil, published around 29 BC. The title of the poem comes from the Greek term for 'agricultural things' (in a loose translation). Although the description of bees and their habits in this case is a fair one, Roman beliefs on the origin of this insect are somewhat less ordinary.

MYTHOLOGY AND HISTORY

The *Geoponica* (you will note the same root in the name as Virgil's poem) is a tenth-century collection of Byzantine agricultural lore that spans some twenty volumes. Within it is found a description of a ritual called 'bugonia'. This stems from a belief that bees were spontaneously formed from within the carcass of a dead cow. The name derives from the Greek for 'ox progeny'. Ancient Greeks sometimes called honey bees *bugenēs*. The *Geoponica* lays out detailed instructions on how to create the bees. If you are not a carnivore, you may want to skip the next quote and read on from the following paragraph:

> Build a house, ten cubits high, with all the sides of equal dimensions, with one door, and four windows, one on each side; put an ox into it, thirty months old, very fat and fleshy; let a number of young men kill him by beating him violently with clubs, so as to mangle both flesh and bones, but taking care not to shed any blood; let all the orifices, mouth, eyes, nose etc. be stopped up with clean and fine

50

linen, impregnated with pitch; let a quantity of thyme be strewed under the reclining animal, and then let windows and doors be closed and covered with a thick coating of clay, to prevent the access of air or wind. After three weeks have passed, let the house be opened, and let light and fresh air get access to it, except from the side from which the wind blows strongest. Eleven days afterwards, you will find the house full of bees, hanging together in clusters, and nothing left of the ox but horns, bones and hair.

One wonders if our love of honey would be so strong if this were the actual way that bees came about!

There was a similar idea that suggested that wasps were created in the corpses of horses. It is possible that this arose from a misreported observation because of the similarity between bees and wasps. We will return to this in a moment, when we look at the actual biological evolution of the bee. There are also species of flies that resemble bees; a fact that may also account for the source of any confusion.

Another suggestion to explain where the concept of bugonia might have emerged from is that bees choose to nest in cavities. It would therefore follow that the body cavity of a dead animal would provide just as much shelter as any other space. This would seem to be unlikely, however, when we note that it has been observed that bees will not settle on corpses.

We find an interesting description of a similar concept to bugonia in the biblical Book of Judges, where Judges 14 tells the story of Samson and the bees. In this tale, Samson kills a lion. When he later returns to the body, a swarm of bees has settled in the carcass and produced honey. Samson collects the honey, and later he presents the event in the form of a riddle:

> Out of the eater, something to eat;
> out of the strong, something sweet

The meaning of the riddle has long since been lost, but it may relate to a saying that honey can be found during Leo's month – that is, when the sun is found in the astrological sign Leo. Many biblical stories have their roots in the older religions – even the name Samson comes from Shamash, the Hebrew for 'day'. In contrast, Samson's wife Delilah derives her name from the Hebrew for 'night', Lilah. This links Samson with Gilgamesh, the Sumerian king, who had Shamash as a patron deity. So, Samson is a sun god, related to both Gilgamesh and Hercules. All three of these characters share a lion-killing story.

Bizarrely, in one of those wonderful pieces of trivia that make the study of folklore so fulfilling, we can bring together all of this history and belief, and link it to a tin of syrup! If you look carefully at a tin of Tate & Lyle Golden Syrup, you will notice between the lettering an illustration of a lion. It looks like it is having a siesta – it isn't. Many people are shocked when they discover this fact. It is most definitely dead, and around its body swarm a number of bees. It seems like a very strange icon to place on a food product, but it has been there for a very long time; record-compilers Guinness confirm that the famous green and white tin is the world's oldest unchanged brand packaging. Abram Lyle had very strong religious beliefs, and so it was he that elected to include the

Traditional beehive. Photo: Tracey Norman

visual representation of the Samson story in order to illustrate a quote to tie in with the product. The phrase 'out of the strong came forth sweetness' appears alongside the beast.

The concept of bugonia emerges from the mythology surrounding Aristaeus, the god of beekeeping. Aristaeus loses all of his bees due to some sickness. Confused and upset at this, he calls upon Proteus for advice. The sea-god reveals that the loss of the swarm is a punishment for Aristaeus's role in the death of Orpheus's wife Eurydice. Aristaeus had developed more than a passing interest in the nymph. Whilst chasing her along a riverbank one day, he caused Eurydice to step on a snake. Alarmed, the animal bit her. Eurydice perished from the wound, leading to the famous story of Orpheus travelling to the underworld to bring his wife back. The punishment that led to the destruction of the bees was meted out by the Nymphae, fellow oak-nymph sisters of Eurydice. Aristaeus was told to perform the bugonia ritual in atonement, an act that ultimately led to new bees forming.

The belief in bees forming in the carcass of a cow also links to the suggestion that the health of a hive could more generally be improved if it was fumigated using cow dung. European books on beekeeping include mentions of the bugonia ritual, even as late as the 1700s.

If the bee did not emerge from slaughtered oxen, then where did they come from? Their story begins 135 million years ago. At this time, plants pollinated using the wind to carry their seeds and spores. The whole process was unpredictable and highly ineffective, and so nature created better ways of dealing with the need. Insects evolved to eat pollen, which is highly nutritious. Plants then evolved brightly coloured petals in order to attract the insects. As they flitted from plant to plant, they would carry traces of the pollinators with them, causing the plants to more successfully reproduce.

The most successful pollen gatherers to emerge were the bees, and the first bees evolved from the wasps that were already thriving on the planet. We might consider the common yellow-and-black-coloured wasps that we see today as being most like bees, but a lot of species are very different to that. Many are parasitic. If you are familiar with the *Alien* film franchise, then you will also be familiar with the life cycle of the parasitoid wasp, as the reproductive methods of the xenomorph alien were based on the insect. This particular strain of wasp lays its eggs in other host insects using a tube, with the grubs eating through from the inside and bursting out. Thankfully (for other insects at least) not all wasp species operate in this way. The Sphecidae family of wasps feed their grubs in a nest, and it is from this strain that the bee evolved.

The process happened when pollen became introduced to the nest, probably carried on the legs of the returning wasps in the first instances. Over time it became apparent that the pollen was much richer in nutrients than the other food being brought into the nest, and evolutionary processes took care of the rest until the bee emerged as a distinct animal. DNA studies suggest that the divergence of bees and wasps took place around 130 million years ago. Bee fossils are very rare, with the oldest bee being encased in amber around 80 million years ago. Since this time, the bee has developed approximately 25,000 known species, with around 250 being what we would now call the bumblebee.

The fat and fluffy bumblebee, buzzing from flower to flower to return the ingredients for making honey to the nest, occurs the most in the folklore of the animal, purely because of the fact that it was the type most suited to being kept for this purpose. Its name in this form didn't emerge until the twentieth century, however – Charles Darwin would have referred to this as the humblebee. This was not because of the mundane nature of its work but simply because as it flew, it hummed.

So how did the name change come about? There are a couple of possibilities. In Beatrix Potter's 1910 story, *The Tale of Mrs Tittlemouse*, the chief character responsible for making nests in the garden is named Babbitty Bumble. Maybe this planted the seed of the name. At a similar time, the apocryphal story emerged that the bee's aerodynamic design meant that it was impossible for it to fly under the known laws of physics. Discussions turned to the bee 'bumbling' around between plants. In the next major publication on bees, in the 1950s, the term humblebee had gone and was replaced by what we now know. As an interesting aside, the archaic term was developed in one other way. Author J.K. Rowling, who often draws on folklore within her *Harry Potter* universe, chose the name Dumbledore for her headmaster because she saw him 'wandering around the castle humming to himself'.

Collecting honey from wild bee colonies is possibly one of the oldest human activities, and one that is still carried on to this day. The earliest recorded evidence comes from rock paintings around 13,000 BC and the practice is continued now within some aboriginal societies.

Beekeeping – the practical management of social species of honeybees – is also very old. Honeybees were kept in Egypt from antiquity. We find inscriptions describing honeybee production in Pabasa's tomb (c.650 BC) and pots marked that they contained honey were discovered in Tutankhamun's tomb. In Egyptian mythology the belief was that bees grew from the tears of the sun god Ra that landed on the desert sand.

In prehistoric Greece, beekeeping was a high-status activity. Archaeologists have found beekeeping paraphernalia at Knossos and, similarly, intact hives have been unearthed at Bronze and Iron Age sites in the Jordan Valley. The Greeks believed that a baby whose lips were touched by a bee would become a great orator or poet. Homer, who fits the description, suggested that Apollo's gift of prophecy came from three bee maidens. We might compare this to the fates (discussed in Chapter 1) and, similarly, interpret it as symbolic of the trinity.

Indeed, within the Christian faith, the bee is symbolic of the soul and the life of the hive is a model of Christian society. Beeswax used to be the only substance suitable for turning into votive candles. We can even find the kinds of charms more readily associated with the older religions being used to try and cause bees to settle near an apiary. Medieval swarm charms all address bees, or the queen, in the name of the Trinity (again, note Homer in the previous paragraph) or the Virgin Mary.[2]

Some argue that such charms were constructed by churchmen. However, examining an Anglo-Saxon charm, it appears more likely that they were adapted from older versions to suit the new religion. This pre-Christian charm roughly translates as:

Throw dirt with the right hand under your feet and say, 'I take under foot, I am trying what earth avails for everything in the world and against malice, and against the mickle tongue of man, and against displeasure.' Then throw some gravel over the swarming bees and beg them to sink into the earth, not to fly to the wood, and to serve the owner well.[3]

The sheer number of swarm charms that can be identified (particularly in Christian form in the thousand-year period between the ninth and nineteenth centuries) suggests the importance of the act of swarming, and the folklore would seem to back this up. Folklore tends to naturally develop around events that are desirous to happen, often in the forms of bringing good or bad luck.

Folklore pertaining to bees is one of the most prominent of all animal folklore. This may be due to the many links that the bee has with both creation stories and key mythology in different cultures. In the creation myth of the Kalahari San people of South Africa, for example, a bee carried a mantis across a river. As they crossed, the bee left the mantis on a floating flower. Before it died from the journey, the bee planted a seed in the body of the mantis and this grew to become the first human. This story is particularly interesting when you consider it alongside the detail of the life cycle of the wasp from which the bee developed, discussed at the beginning of this chapter.

The bee was often seen as sacred in Eastern mythology, because it formed a bridge between the physical world and the underworld. There are some similar parallels in older

Western beliefs too. The Irish Goddess Brigid held bees as sacred because her hives brought nectar from the apple orchards in the Otherworld. In catholic Christianity, Saint Gobnait (suggested as having lived in approximately the sixth century) is most likely a figure who arose from the older Brigid stories. Gobnait is the patron saint of bees and beekeeping, and there are a couple of different stories surrounding her and her work with the bees at the convent of which she was abbess. In one, she cures a nun with honey from the abbey hives. Another story tells how Gobnait employed the bees to prevent cattle thieves from stealing livestock. Like all good folk tales, there are multiple versions. One variant has the bees turn into soldiers and the hive into a bronze helmet. Another has the hive turn into a bell, which becomes one of Gobnait's symbols. And in yet another, the bees attack the invaders in their animal form, employing their own weapon of the sting.

In the Roman pantheon, Jupiter was said to have given the bee its sting, as a way of protecting the honey that it produced. His wife Juno, however, stated that there must be some form of payment for a gift that may be used as a weapon, and this was the reason for the bee dying if it used its sting. In fact, the sting genuinely is a protective mechanism against honey theft, as the honey bee is the only variety that dies after stinging. And even then, this only happens if the bee stings a mammal. A bee will easily survive stinging another insect. This is due to the bee sting having a barbed structure designed to penetrate further into the enemy. In the case of mammals, the skin is very fibrous and this leads to the sting and part of the abdomen being ripped out.

FOLK MEDICINE

As with many things, of course, where there is a bad side there is often a positive too (although not necessarily for the poor bee). We can find bee stings and bee venom used in various ways in traditional folk medicine. One cannot say for certain just how effective or otherwise some elements of folk remedies might be, but it has often proven to be the case that 'old wives' tales' have more than a grain of truth behind them.

Take, for example, the idea that copper coins might be effective in easing the pain from a bee sting. Claims of this nature have been circulating, on the internet at least, for well over ten years. One version of the story tells how a woman working in the garden gets stung and has a bad reaction, causing her arm to swell. She goes to a clinic and gets anti-histamine and cream, but it is not effective and the following day she sees her regular doctor, who treats the infection with antibiotics. During the treatment, the doctor advises that if it happens again, she should put a penny on the wound for fifteen minutes. Conveniently, she (or a relative) gets stung again soon after and applies a coin, and this time there is no swelling or infection.

As with many urban folk tales, we cannot tell if the event truly happened or not. But what about the theory? There isn't any scientific proof to say that applying a copper coin to a sting is effective. Whilst there are medical creams that contain copper mixtures within the ingredients, they would be more carefully balanced than the metal content of a coin! It is also worth considering that since the early 1980s, most

Bees searching for pollen. Photo: Tracey Norman

'copper' coins actually contain very little copper. All of that is not to say that copper coins are not used at all within traditional folk cures. In fact, they are professed to be good treatments for various problems, but these are not usually connected to stings. Many people will at least be aware of, or possibly even wear, copper bracelets as an aid to easing the pain associated with rheumatism. Aside from bracelets, it is also said that a penny should be placed in the shoe for the same reason.

Some plants may be effective in helping to cure bee stings, if the folklore of various areas is to be believed. Renowned plant folklorist Roy Vickery has identified variously chickweed, houseleek and sliced onion as being used for this effect.[5]

The venom in a bee sting has also been put forward as a curative. The Chinese have used it for many years, but more recently bee sting treatment has been offered as an anti-inflammatory. There are many warnings to be found about this method, as there is no proven evidence for its efficacy and, of course, allergies to bee stings can lead to death in extreme cases, so it is not advised by commonplace medical authorities.

George D. Hendricks published an article in the journal *Western Folklore* in 1975[6], in which he listed many examples of folk superstitions that he had collected from the Texan and South-western areas of the United States. Among these was a drink believed to cure a baby of croup (a condition that can affect the airways of young children):

> Catch and boil three bees, and then feed the baby this 'Three Bee Tea'. An old man on our farm, hearing the screams of his sick grandchild, shouted, 'Catch me three bees! Get water to boiling!' My mother reached the baby first and took the child from its mothers arms, only to have it look up at her and laugh. Even though the child recovered from the croup, the three bee tea was administered to it.

That particular example was recorded in Leesville, Louisiana.

HONEY

It is, of course, more common to focus on the honey that the bees produce rather than the bees themselves when looking towards medical properties. Many people still hold that eating a teaspoon of honey produced locally to the area in which they live will ease the symptoms of hay fever. A traditional Scots recipe of heather honey, cream and whisky in equal parts was used to try and cure wasting diseases. The idea of feeding milk and honey to infants comes from an older practice of giving them hazel milk and honey. More recently, we use milks such as hazel or almond as a dairy alternative and we also find many shops stocking milk and honey products as a skin treatment.

CUSTOMS AND SUPERSTITIONS

The collecting of honey and the keeping of bees has always been a very important undertaking. We have already established how far back in history we can find practical examples of beekeeping; for this reason, we naturally find a plethora of bee superstitions in the folklore record, some of which are still followed to this day. It is always better to try and ensure good luck than take a risk, after all.

Possibly in part because of the ideas of the beehive as being reflective of society, some superstitions relate to the importance of treating bees as part of the family. The most well-known of these is probably the practice of 'Telling the Bees' of the death of a loved one, or other significant event in one's life. Because it is usually associated with a family death, some think that 'telling the bees' may come from the Celtic ideas already discussed surrounding bees and the soul. It isn't possible to make this link for certain, but we can say that the idea became most common during the eighteenth and nineteenth centuries. We find evidence of it happening in both America and Europe – it was well recorded in New England, and so may have travelled to America with the settlers. Quaker poet John Greenleaf Whittier titled a poem written in 1858, 'Telling the Bees', and this gave a good example of the practice.

Imparting the news of a death to the bees was done on a hive-by-hive basis. The beekeeper would first knock on the hive before telling the news of the death, and the hive would also be covered by a black cloth or similar piece of material for the period of mourning. In some examples, the bees were sung to rather than spoken to. Author Tammy Horn, who keeps bees herself, notes that in New Hampshire at least, the verses of the song must also rhyme.[7] She also notes another custom relating to death. When a family suffers bereavement, the eldest son should move the hives to show that a change has taken place. In some cases they are moved to the right; in others they are turned to face the door of the family home. This ritual was known as 'ricking'.

We still find rituals and customs such as 'telling the bees' taking place to this day. As recently as the last fifty-or-so years, we find people suggesting that these practices still take place in rural areas because of firm beliefs, but it is often more likely that they are done out of a sense of tradition. In this case, noted from a British newspaper in the middle of the twentieth century by folklorist E.F. Coote Lake, both are mentioned:

A man went up to a beehive in a country garden yesterday, knocked three times, and said: 'The master is dead. The master is dead.' The master was Mr Ernest Langley, of Burton

Latimer, Northants, who died in hospital. Before he died, he asked his friend Mr J.C. Lutener to tell his bees of his death – an ancient custom among beekeepers. Mr Lutener, chairman of the Kettering Beekeepers' Association, of which Mr Langley was a member, said: 'Mr Langley firmly believed in the custom, and said that if the bees were not told they would leave the hive … Mr Langley was very much a man of the country and liked to keep up country customs. I was glad to do it because it was his wish.[8]

It was thought to be extremely bad luck to ignore the bees and not tell them of a death or other big family event. To not do so would mean that the bees would either depart the hive, or perish altogether. One story from the end of the nineteenth century tells of a man from Norfolk who purchased a hive of bees at an auction. Upon getting them home and finding them looking very weak, he suspected that they had not been allowed to mourn the death of their former owner. The story says that after draping the hive in black for a period of time, the bees recovered. Was this the problem or not? Even within the spheres of folklore there could be other reasons why the bees suffered problems. The man in question had bought the hive at auction, yet other beliefs say that a hive will not produce honey if you pay for the swarm. Payment for bees should be made in honey and comb, rather than in money, or by exchanging another commodity of equal value.

Bees are also said to not thrive as part of a family who have lots of arguments, as the animals are peaceful creatures and do not like either the atmosphere or the noise in such a situation, leading to them leaving the hive. For similar reasons, bees are said to have a dislike of swearing. Both of these ideas are found in Britain and in the US.

In some areas, if the bees do leave a hive, then another ritual called 'ringing' or 'tanging' must take place. Generally, a key was hit against a shovel to make a noise that made it clear to neighbours that the owner did not have to respect property boundaries when pursuing the swarm, and hence could follow it wherever was necessary to retrieve it. The term 'tanging' derives from the practice of rough music that sometimes replaced bells at a wedding.

Singing to bees in their hive is not found solely in relation to death customs, but may also be used as a means to try and encourage production. This often draws on the Christianised or religious elements of the swarm charms mentioned earlier in this chapter. In both England and Germany, for example, hymns or psalms would be sung to the bees in the apiary to make them produce more honey. Sometimes, the bees might even try to sing themselves ... in a way. In Christian societies it was said that the bees would hum loudly on Christmas Day in celebration of the birth of the saviour.

In some Catholic areas, particularly in America, the sacred wafer was placed near the beehives in order to try and increase the honey production. In Cornwall, however, which has much stronger Celtic roots, this was considered to be sacrilegious. The county did have some Christianised customs relating to bees, however. Amongst these was the belief that the animals could only be moved on Good Friday.

If bees did swarm, then this would carry its own set of beliefs with it. These beliefs were often contradictory, depending on the area where they were recorded. For example, some lore in Wales notes that it is unlucky if a swarm entered your house, as this was thought to presage death. However, other lore records that it is lucky if it enters, but unlucky if it leaves. In some areas it was said to be a bad omen if a swarm settled on a dead branch. This would bring death for the witness, or for the family of the beekeeper. In other places the same portent was true if the bees landed on the roof of a house. In still other examples, bees nesting on a house prevented marriage.

When it comes to not getting stung, however, the prevention of marriage may have had its advantages, as it was said that bees responded very favourably to chastity! It used to be believed in some areas that a girl who could walk through a swarm of bees without getting stung was protected because she was a virgin. It was also thought that bees would not sting babies.

In both England and America, we find other beliefs relating to who may or may not get harmed by bees. These all seem to tie in with the ideas of the bee liking peace and harmony. In Maryland, it was recorded that bees would not harm anyone who showed a good disposition. And in England, it was recorded in William Ellis' 1750 book *Modern Husbandman* that:

> All that keep bees should love them, for these hate those that hate them. A farmer's wife loved them very much, but her husband hated them; they would sting him, but not her.[9]

Ellis came from the county of Hertfordshire, where he farmed at Little Gaddesden, and so it is safe to assume that this and the many other pieces of folklore and proverbs that he recorded would have been local to his area. This is true of the case of a well-known proverb:

A swarm of bees in May is worth a load of hay;
But a swarm in July is not worth a fly

which Ellis expands upon by adding a local variant from his area:

A swarm of bees in May
Is worth a cow and her calf, and a load of hay[10]

This proverb is not only well known, but also well travelled, being found as far apart as Dorset in England and New York State in America. As noted earlier, it would most likely have travelled there with settlers and, in both cases, it also includes extra lines for the month of June:

A swarm of bees in June
Is worth a silver spoon[11]

On 3 March 1961, a very strange swarming incident was recorded in an article in the *Shrewsbury Chronicle*. Folklore collector E.F. Coote Lake again recorded this in an article for *Folklore*, the journal of the Folklore Society:

Sam Rogers, who died suddenly six weeks after his retirement as Myddle's postman, a job he held for forty-one years, was devoted to his bees. At his death his children carried out the old traditional custom of walking round the fourteen hives 'telling the bees', to stop them, as legend has it, flying away. On the day after his funeral a memorial service was being held in the church when it was noticed that swarm after swarm of bees were coming from the direction of Sam Roger's home in Lower Road, Myddle, a mile and a half away as the crow flies. They settled in a great swarm all over the flowers on the grave to the astonishment of the congregation when they came out from the service. They thought it fantastic.

The rector, the Rev J.C. Ayling told the Chronicle 'It was a remarkable sight. The bees were clustered all over the grave. The only logical reason, I suppose, is that the flowers on the grave attracted them.' But experts are puzzled by this behaviour, because at this time of year bees are still sluggish and rarely fly from the hives.

The chairman of the Shropshire Beekeepers' Association, Mr H.D. Pocock, said it was unusual for bees to fly such a distance at this time of year, he had never heard of anything like it before – perhaps the flowers on the grave attracted them. But it was very odd; perhaps it was a result of the children 'telling the bees'.

By nightfall the bees had all flown back to their hives in the cottage garden where Sam Rogers had looked after them for so many years. Mrs Lilian Hayward of Church Stretton, Shropshire, had three letters from correspondents bearing out the newspaper item.[12]

CULTURAL IMPORTANCE OF BEES

It is of little surprise that bees and the keeping of them have always been so important to us as a race. Until the fifteenth or sixteenth centuries, the honey that they produced was the only sweetening agent that humans possessed. And so these little creatures have always been embedded in our culture in so many ways. The Welsh bards called Britain the 'Isle of Honey' because of the large numbers of wild bees that they saw in the forests.

Honey was known as 'the nectar of the gods' and was used in many ways as an offering, because it was considered sacred. Jewish children beginning education would have honey dripped onto the first page of their book to ensure that they saw their learning as sweet. At New Year also, Jewish people ate apples dipped in honey. Again, this represented a sweet year ahead.

In symbolic terms, we might also find honey in connection with fertility. After all, the time that follows on from the marriage ceremony is called the honeymoon. At one time, a hot drink made from honey, milk and wine would be given on the night of the honeymoon. Couples would also traditionally drink mead, which has honey as a base. In Croatia, honey would be put on the threshold of a house before a new bride and groom entered. At Balkan weddings, the faces of the couple getting married would be painted with honey. These customs were all related to fruitfulness.

Celtic myths held that bees had great wisdom, highlighting their importance, and they were described as messengers between gods and mortals. In later Christian myths, this developed from the bee moving between realms to stories that it came directly from Paradise. In the Western Islands of Scotland, these ideas continued and it was said that the bees had the ancient knowledge of the Druids. A saying recorded in this area tells one to 'ask the wild bee for what the druid knew'. In the Highland areas, people used to believe that the soul would leave the body while someone slept, or while they were in a trance, and would travel in the form of a bee.

Across the world, we find pieces of folklore relating to bees and beekeeping. And as we discussed here, there are many similarities between them. Sometimes this is due to stories travelling as people moved around the world, but often the roots seem to be much older, as if at the subconscious level we all retain a certain amount of collective folk memory. These memories resonate with us, without us knowing why, and make us human on one level. This is demonstrated in an old ghost story from Lithuania. Although it contains many themes that we have already discussed, and that also appear in older stories from the country, the tale as it was recorded was absolutely believed as true by the teller. The story was collected in the summer of 1949, by Jonas Balys in Pittsburgh, USA, and was published three years later in the journal Midwest Folklore.[13]

The lady who told the story was Mrs Magdalena Takazhauskas, who was born in a village called Versnupiai in western Lithuania in 1877. It concerns a beekeeper named Kurauskas, who was generally considered by everyone who knew him to be a good man, but who once undertook what was considered to be a sacrilegious act.

Der Bienenfreund (The bee friend). Oil painting by Hans Thoma, 1863/4. Depicts a man imparting news to the bees. (Creative commons licence)

One day, while attending Holy Communion, Kurauskas took the sacred wafer from his mouth and smuggled it home, where he placed it on one of the beehives in his care. This caused the bees to become very productive, and the estate owner for whom Kurauskas worked was very pleased with him.

Sometime after, while very ill, Kurauskas wandered outside and drowned in a swamp that had grown up in the yard following a period of heavy rainfall. Three days later, after the body had been buried, mourners gathered at the funeral feast saw the ghost of Kurauskas standing watching them and ran away terrified.

What followed from this event was a series of hauntings, which did not seem to befit the good reputation that Kurauskas had held in life. His ghost would appear to men whom he had enjoyed wrestling matches with when he was alive, and they would be forced to wrestle him all night. One was said to have died from exhaustion because of this. He appeared to a Jewish man, to whom his widow had sold his suit, and stole it back. On one occasion, his spirit hung a horse from the beam of a barn by its bridle and the animal had to be cut down before it suffocated.

The local priest who investigated the haunting learned that Kurauskas had been a successful beekeeper, and so called on the help of another of that profession to look into the events. This man examined the hives and discovered that in one of them, the bees seemed to be continually buzzing the melody of a popular hymn. The beekeeper therefore opened the hive and inside he found that the bees had created a chalice from their wax. Examining the chalice, the man found the sacred wafer inside.

From here, the story takes a more vampiric turn as the grave is exhumed and the man is discovered to be a revenant. His haunting is stopped using methods that are common in European vampire lore, and the body being staked.

The story of a beekeeper who puts a consecrated host into his hive is common in this part of the world and always leads to productive bees, followed by some form of death and haunting as retribution for the stealing of the sacrament. But it is interesting in this case to see the other pieces of folklore, such as the bees singing hymns, which we have noted from Western traditions.

Bees are undoubtedly important in folk terms, but as we have heard so many times recently, they are also vital to the well-being of our planet. This idea is summarised perfectly by authors Benjamin and McCallum in their book *A World Without Bees*, where they note that:

Bees are the canary in the coalmine of the Earth.

So, if a bee should land on your hand, don't hurt it or shoo it away. Not only are you doing your bit for our ecological well-being but, according to old folk beliefs, money will come your way.

Chapter Two Sources

1. Virgil (70BC – 19BC) 'The Bees', Georgics Book IV
2. Fife, A.E., 'Christian Swarm Charms from the Ninth to the Nineteenth Centuries', The Journal of American Folklore, Vol 77, No 304, American Folklore Society
3. ibid
4. Hendricks, G.D., 'Dishrags, Dogs and Doors', Western Folklore Vol 34 No 1, Western States Folklore Society
5. Vickery, R., 'Vickery's Folk Flora: An A-Z of the Folklore and Uses of British and Irish Plants', 2019, W&N
6. Hendricks, G.D., 'Dishrags, Dogs and Doors', Western Folklore Vol 34 No 1, Western States Folklore Society
7. Horn, T., 'Bees in America: How the Honey Bee Shaped a Nation', 2006, University Press of Kentucky
8. Sunday Express, London, December 4 1955
9. Ellis, W., 'The Modern Husbandman, or the Practice of Farming', Vol 4 Part 2 1750
10. Ellis, W., 'The Modern Husbandman, or the Practice of Farming', Vol 3 Part 1 1750
11. New York Folklore Quarterly, VII (1951), 23, Cf Brown Collection, VII, 435, No 7515
12. Coote Lake, E.F., 'Folk Life and Traditions', Folklore, Vol 72, No 2 (Jun 1961), Taylor & Francis Ltd on behalf of Folklore Enterprises
13. Midwest Folklore, Vol 2, No 1 (Spring 1952)

3

BLACKSMITHS AND METALWORKING

Under a spreading chestnut tree
The village smithy stands;
The smith, a mighty man is he,
With large and sinewy hands;
And the muscles of his brawny arms
Are strong as iron bands.

His hair is crisp, and black, and long,
His face is like the tan;
His brow is wet with honest sweat,
He earns whate'er he can,
And looks the whole world in the face,
For he owes not any man.

These words are taken from the first two stanzas of the poem 'The Village Blacksmith' by American poet Henry Wadsworth Longfellow. First published in 1840, it paints a respectful and, at the same time, stereotypical picture of the blacksmith as a powerful figure at the centre of a village community.

Why was it that the blacksmith was so respected years ago? As we shall see, it was partly due to the necessity of his (most smiths were men) trade to so many people. But also, as with many very skilled professionals, the abilities of the blacksmith presented him with almost supernatural powers. These powers were forged (no pun intended) in history through the mythology surrounding the trade.

MYTHOLOGY AND HISTORY

Evidence of early metalworking can be found dating from way before the Iron Age. In fact, the association of iron with forms of magic has early origins too. One of the oldest iron items found was a dagger discovered during excavations in the tomb of Tutankhamun in Egypt. Its construction was placed at around 3,200 BC, which means that it was made around two millennia before the Iron Age.

In this case, the metal was not fashioned in the normal way, but was a gift from the gods. X-ray spectroscopy tests carried out on the dagger in 2016 showed that the composition of the metal was in line with that found in iron meteorites, which have fallen to earth for billions of years. It was most likely the sort of metal that was worked by Egyptian craftsmen to form the blade of the dagger.

The Egyptian creator god was named Ptah. His deity originated in the first dynasty, centred around the Egyptian capital of Memphis, before his cult expanded across the rest of the country. Because of his role as a maker, Ptah became the patron of craftsmen at that time. The Greeks identified the figure of Ptah from the Egyptian religion with their own Hephaestus, who was blacksmith to the gods. We find references to Hephaestus in Minoan scripts, which certainly points to an early origin and may also suggest contact between the Egyptians and the Greeks. The latter referred to amber as 'the solidified tears of the sun', and this is a phrase that has also been noted associated with the meteorite metalworkings of the Egyptians.

The Hephaestus of Greek myth may be found as Vulcan in the Latin. In works of art, both figures are often depicted with hammer and tongs, common tools of the blacksmith. Through early trading along common routes, these figures probably both drew from and informed those in other cultures, and we certainly find many important blacksmiths, both mythological and real world, in other societies. The Hindu religion aligned their own deity, Tvastar, with the Roman Vulcan. Also a creator god, Tvastar was depicted as a carpenter, although a fire device and a ladle were among his attributes. Again, he was blacksmith to the Devas (the other Hindu deities). Other examples from mythology include Ilmarinen, a Finnish blacksmith known as The Eternal Hammerer who may be found in the epic poem the Kalevala, and Ogun, god of iron, a spirit of metalwork from Nigeria.

Tribal smiths in parts of Africa are still looked upon with respect in different ways. Kenyan blacksmiths are seen very much in the way that other tribes look upon witch doctors; being imbued with magical powers from the spirits. Much like witches in the early modern period, their abilities and skills were needed and yet they were driven to the fringes of a community and were forced to live in isolation for protection. In Ogun's country of Nigeria, smiths are seen as a form of priest or guide in spiritual matters, and are also viewed as healers. The ability to heal being connected to a blacksmith or forge is something that we will return to later in the chapter. Here too, there are many traditional dances and rituals associated with the work of the smith, and these bring the whole community into one place to take part. Other cultures reflect both of these aspects; of spiritual power and of a sense of ritual. The Yakut ethnic group of the Russian Federation have a proverb, which reads that:

The blacksmith and the shaman are from the same nest.[1]

But interestingly, they accord more supernatural power with the blacksmith than they do with the shaman. Their stories say that the shaman cannot kill the blacksmith, because at each attempt the latter's tools turn into winged creatures that attack the former. Novice Yakut blacksmiths undergo complex initiation rites as part of their journey to the craft, which were recorded by A. Popov in an article for the *Journal of American Folklore*.[2]

Blacksmiths are, of course, found in the more prominent of the older mythologies such as the Celtic and the Norse. Völundr, more commonly known in the Germanic mythological sources as Weyland the Smith, was a master blacksmith

Vulcan forging the Thunderbolts of Jupiter. Oil painting by Peter Paul Rubens, 1636-38. (Creative commons licence)

who used his supernatural skills to fashion magical weapons, armour and ornaments. His story may be found in the thirteenth-century Icelandic text, the *Poetic Edda*. It tells of his capture by the evil Swedish king Nidud, who cuts Weyland's tendons, making him lame and therefore unable to escape. Placed on a remote island, the smith nevertheless manages to leave by making himself either magical wings or a feather robe, depending on the version of the story that you read. Many people would probably associate this character with a famous English Heritage site, Wayland's Smithy in south Oxfordshire. There are often variations in spelling between Weyland and Wayland, such that the two are almost interchangeable in modern writing. This is a two-phase Neolithic tomb, the first part of which was built *circa* 3500 BC. In fact, there is no association at all with the Germanic mythology, and the name was first linked to the barrow around 955 AD in a Saxon charter. It is said, in what is described on the English Heritage website as a 'local tradition',[3] that if you leave your horse tethered to a certain spot at the site, then Wayland would magically re-shoe it while you were away.

There are two roots that could be ascribed to this story. Firstly, there are long-established traditions within folklore that hidden folk such as fairies will undertake tasks in return for offerings. In some legends, Weyland is known as King of the Elves, and so maybe there is some old connection to be found between the two. The second possible explanation is less supernatural. It is possible that at one time there were two communities who lived close to the area. The smith, with the skills for shoeing the horses, lived in one community. Members of the other needed his services but the two rival communities would not be seen to be dealing with each other and hence the business transaction was made in this way.

It is interesting to note that among the many blacksmith-related superstitions, there is one that says that you should place a penny on the floor if you are using somebody else's fire or tools. This is because it is not honourable to do so without making a payment to the person from whom you are borrowing, but at the same time it is also not gracious to accept a payment. Leaving a coin on the floor overcomes both of these problems. The idea of leaving a coin at the site seems to have perpetuated, as in modern times people have taken to leaving coins as offerings in cracks in the stones. Although well meaning, these sorts of practices put ancient sites at risk, and so custodians of Wayland's Smithy regularly remove the coins, which are donated to charity. The proliferation of the ritual of hammering coins into tree trunks to form coin trees is similarly problematic.

Irish mythology has Goibniu (who is known as Gofannon in the parallel Welsh stories). Most mythical smiths have stories attached to them wherein figures of power or authority request their services to forge special weapons, and Goibniu was no exception. He even had the distinction of making a metal prosthetic limb for a warrior named Nuada, whose arm was lost in a battle.

Goibniu's name went on to form the old Irish word for 'smith', and the importance of the role in the community led to the widespread occurrences of the name Smith and its equivalents in different languages. The actual linguistic root of smith is contested, with suggestions that it may come from the Old English term 'smythe', which means to strike, or from 'smithaz', a Proto-German word for a skilled worker. The 'black' element of the term would certainly seem to come from the oxides that form on the metal whilst it is being heated, and therefore also coat the blacksmith and all of the equipment being used in the craft.

As with much ancient mythology, we find equivalence within later biblical texts and readings. The Torah notes Tubal-Cain as being the first smith, or the 'inventor' of the art of metalworking. But even here, there are some doubts in the origin story. While some biblical scholars believe this translation to be accurate, others hold that the actual translation from the original Hebrew should be Tubal. They think that the 'Cain' (or qayin) was added later, and more properly translates as 'smith'. There was an Assyrian race known as the Tubal, whose principal manufacture was bronze and iron vessels. The man Tubal in the biblical texts may therefore have been a kind of representative ancestor for the race as a whole.

As noted earlier, any tradesperson who was highly skilled had some kind of supernatural element ascribed to them as a way of explaining why they were so much better at their art than others. This was a way of making sure that other people did not feel a sense of inferiority about themselves. The blacksmith is no exception, and in many cultures we find stories explaining how a deal was struck with the devil in return for granting these special powers of the trade.

THE BLACKSMITH AND THE DEVIL

Work carried out for a 2016 article[4] traced population and language changes over time, and showed that it is most likely that the story *The Blacksmith and the Devil* is in fact the oldest known folk tale. It crops up at different times, and in different cultures, with variances to adapt it to the circumstances in which it is found, but the main elements remain the same: a blacksmith meets the devil and traps him in some way, only agreeing to free him in return for the secrets of metalwork and forging. The story has been collected from such diverse places as Russia, Scandinavia, India and America, and of course the Grimm Brothers included it in early volumes of *Children's and Household Tales*.

In one version, the blacksmith is approached by a woman who tries to distract him from his work. However, the wise smith notices that the woman has cloven hooves instead of feet and so realises that it is the devil. He grabs the devil by the nose using his tongs and removes him from the smithy. The next day, the smith has another visitor in the form of a wealthy businessman who tries to offer large sums of money in exchange for some deal. Again though, the devil has not bothered to cover his feet – he never seems to learn in these tales – and his hooves are on show. In a repeat of the previous day, the blacksmith uses his tongs to hold the devil. This time, he will not release him until the devil agrees to impart the secrets of the smiths' trade.

After gaining his wisdom through these supernatural means, the blacksmith then releases the devil – but he also makes him promise that he will not tempt any homeowner who hangs one of the smith's horseshoes over his door, leading to the well-known custom that we still see today.

The story of St Dunstan gave rise to a folk rhyme on the topic:

St Dunstan, as the story goes
Once pull'd the devil by the nose
With red-hot tongs, which made him roar
That he was heard three miles or more

In later Christianised versions of this story, the event is ascribed as taking place in the blacksmith's shop of St Dunstan. One-time Archbishop of Glastonbury and Anglo-Saxon advisor, Dunstan was born near Glastonbury and died in 988 AD. In this particular version the saint recognises the devil as soon as he walks into the shop, although he pretends not to. The devil requests that his horse is shod but St Dunstan, after fetching the shoes, nails them to the devil rather than to the horse. They cause a lot of pain but again Dunstan says that he will only remove them if the devil promises never to enter a house with a horseshoe on the door. In the Christian version of the story, of course, there is no mention of the exchange of supernatural knowledge for skills.

The smith's tongs remain one of the symbols of St Dunstan, alongside a dove. The idea of using the tongs against a demonic creature is extended further in a German folk tale collected in Canada known as 'The Blacksmith and Beelzebub's Imps'.[5] In this story, the blacksmith has sold his soul to the devil in exchange for power over any person who either picks nails from his shoeing-box, sits in a particular chair or climbs a tall pear tree in the garden. Later, the greedy smith sells himself again, this time to the Prince of all the devils, Beelzebub, in return for riches. The arrangement is that Beelzebub can claim the smith after twenty years have elapsed. Fast forward those twenty years, and Beelzebub sends an imp to claim the blacksmith. The smith asks the imp if he would help him finish his business as he is busy, by picking all of the bent nails from his box. As soon as the imp reaches in he loses his power, because of the smith's earlier bargain. The imp is sent back to his master after the smith pinches him with hot tongs from the fire. A second imp is removed in a similar way, after the smith asks him to wait while he washes for dinner; the imp using the only chair available, which is the enchanted one. A third and final imp is sent to claim the soul of the blacksmith.

This time, the smith asks the imp if he will pick the pears from the top of the very tall tree so that they do not spoil, as he is too busy and his wife and daughter cannot climb. The imp is so keen to complete the task for Beelzebub that he helps the smith, with the same consequences. This time, the blacksmith and his apprentice use long iron rods to torture the creature. Beelzebub fails to find any more assistants willing to go for the blacksmith's soul and so, in this story, the blacksmith wins out over the Prince of devils.

THE IMPORTANCE OF THE SMITH

Moving on from the tenth-century figure of St Dunstan, in the medieval period the art of smithing was one of the *Artes Mechanicae* – the seven mechanical arts. These were the seven skills seen as being vital to a community. The others were tailoring/weaving, agriculture, architecture, warfare/hunting, trade and cooking. A legend told by an old Sussex blacksmith and included in the journal *Folklore* in 1884[6] uses the same idea but speaks of the seven trades: blacksmith, tailor, baker, cobbler, carpenter, butcher and mason:

> On the 17th March, AD 871, when good King Alfred ruled this land, he called together all the trades (seven in number) and declared his intention of making that trades-man King over all the trades who could best get on without the help of all the others for the longest period. He proclaimed a banquet to which he invited a representative from each trade, and made it a condition that each should bring a specimen of his work, with the tools he used in working it.

The King decides that the beauty of the tailor's work surpasses all of the others and so he is declared to be suitable to represent the most important of the trades. This decision makes the blacksmith very jealous and so he says that he will no longer work whilst the tailor is so elevated.

In time, all of the others find themselves in need of the blacksmith. King Alfred needs his horse shod and the others all break or wear out their tools. The blacksmith, however, is nowhere to be found. The others all try to use his forge themselves, but the results are far from good.

Eventually, St Clement arrives with the blacksmith. The King apologises for his misjudgement, saying, 'I have made a grave mistake in allowing my judgment in this important matter to be governed by the gaudy colour and stylish cut of the tailor's coat.' All of the other tradesmen admit that they need the blacksmith to mend their items and so the King acknowledges that the blacksmith should be seen as head of all the trades for all time. A feast takes place with much celebration and songs are sung, one of which is known as *The Jolly Blacksmith* and begins:

Here's a health to the jolly Blacksmith
The best of all fellows,
Who works at his anvil
While the boy blows the bellows;
For it makes his bright hammer to rise and to fall,
Says the Old Cole to the Young Cole and the Old Cole of all.

Saint Clement's Day is traditionally celebrated on 23 November. Pope Clement I is the patron saint of metalworkers and blacksmiths, and so they would have a holiday on this date to celebrate his feast day. This is also traditionally known as Old Clem's Night. A house-visiting tradition took place in Sussex on St Clement's Day, which has died out in recent times – although there was a revival in Hastings in 2012. Known as Clemmening, or Catterning, the custom involved calling house to house, asking for apples or beer. Sometimes an effigy would be created with a wig, beard

Blacksmith at work. Photo: Tracey Norman

and pipe, and the figure would be placed at the door of the inn where all of the blacksmiths gathered to celebrate their patron. The alternative term Catterning comes from St Catherine, a virgin saint whose feast day is two days later on 25 November. At the feasts, it was usual for the oldest blacksmith in the area to assume the chair, with the vice-chair position being given conversely to the youngest. Toasts and songs would be given at the meal, which reference some of those mythological aspects that we have already noted. The first toast was to Vulcan:

> Here's to old Vulcan, as bold as a lion,
> A large shop and no iron,
> A big hearth and no coal,
> And a large pair of bellowses full of holes.

This toast would be followed by singing *The Jolly Blacksmith*, another toast and then a song referencing the biblical smith:

> Tubal Cain our ancient father
> Sought the earth for iron and ore,
> More precious than the glittering gold,
> Be it ever so great a store.[7]

With a wide mythological background and a variety of folk tales explaining how their skills came from supernatural roots, we can see how there was often something of an air of mystery surrounding the blacksmith. They would often add to this themselves by only giving secrets of their work to their chosen apprentices. These were often members of their own family and this helps to explain why Smith is the most common family name in Britain.

The surname is frequently found in other countries too, with linguistic equivalents. In France and Spain, for example, names such as Le Fèvre or Fernández come from the Latin word for iron – *ferrum*.

FOLKLORE AND SUPERSTITION

Naturally, a great deal of superstition has arisen surrounding the blacksmith and his trade, as you would expect from an important figure at the centre of a village community like this – and it extends far beyond the lucky horseshoe. There is more, however, to the horseshoe than just hanging it up.

It has often been a source of debate as to which way up to hang a horseshoe for luck. Some people think that if you put it upside down then the luck will run out, for example, but others think otherwise. It is most common at home to hang your horseshoe with the heels facing upward, to keep the luck in. Blacksmiths, traditionally, would do the reverse and hang a horseshoe heels down. But they would hang it over the anvil so that the luck poured onto that vital part of their equipment.

Whichever way up you hang it, it is important that the shoe is new. The luck is believed to be lost from a horseshoe if the horse has thrown it. For the same reason it was said to be bad luck to use horseshoes from the feet of a dead horse on another animal. If you cannot get a new shoe, then you can use an old shoe instead, provided it was removed by the blacksmith himself. If it was done properly by the smith, then the luck would have stayed in the shoe.

Speaking of luck, it was considered to be good fortune in Scotland to be married by a blacksmith. The border town of Gretna Green became synonymous for the ceremonies, which were known as anvil weddings and seen as being slightly disreputable. It was common, particularly when parents were not enamoured of your choice of romantic partner, to suggest that you were going to 'elope to Gretna Green'. It was also said that if a couple were married by a blacksmith, then their marriage will be very happy.

If your skills were bestowed upon you by the devil then, like in the earlier story of the Blacksmith and the Imps, it is reasonable to assume you would want to make sure that he did not come and visit you again. When the fire is

Horseshoe mounted on wall for luck. Photo: Alan Levine www.flickr.com/ photos/37996646802@N01

extinguished at the end of the work day, the long-handled fire tools should be placed over the fire pot with their handles crossed in order to keep the devil at bay. Another practice said to ward off the devil is 'ringing the anvil' – that is, hitting the anvil with the hammer. Why this keeps him away comes from an old folk tale that in many ways is very similar to the story of St Dunstan.

One day, the tale goes, a village smith was shoeing a horse. The smithy was in the open air and the horse was tethered nearby. The devil, walking past, heard the ringing of the anvil and wondered what was going on. He went to investigate and watched as the smith went about his work, trimming the horse's hooves and nailing the shoes into place.

The horse was so pleased with the new shoes that he danced all the way down the road when his owner rode him off.

The devil saw this and decided that his own cloven feet should have a pair of shoes. He asked the blacksmith to fashion a pair, but the man realised that this was the devil.

He made the shoes too small, the devil's hooves were trimmed too much and the shoes were clinched heavily so that they were difficult to take off. The devil left in great pain, which continued for many days afterwards. And this is the reason why, when the devil hears an anvil ringing, he will not go anywhere near it.[8] Ringing the anvil does have a more practical explanation, being a way of keeping heat in the hammer.

An old superstition tells that a smith who, over a period of nine days of fasting, rises each day before the sun, goes to his forge and strikes the anvil four or five times, will have any wish granted to him. There are omens that relate to the time at work also. It was said to be unlucky to shoe the first horse on a Monday on trust, and so the smith would always ask for cash or at least a part-payment when undertaking this job. Also, if the first horse brought to the blacksmith in the morning is poor in some way – thin or ill, for example – then this is bad sign for the smith's fortune. Healthy, bay or dapple-grey horses were definitely welcomed at this time though.

As most blacksmiths were men, a belief arose that it was unlucky to have a woman come into a forge. It was also usually unlucky to steal anything from a forge. In some cases, however, it was actually desirable to do so. It was said that the water in the slack tub in a forge had the ability to cure illness, but the water was only curative if it was stolen. There may be a much older link that led to this belief: the ancient Greeks thought that drinking the water that the blacksmith used to cool tools would prevent pregnancy.

It was once thought, in fact, that blacksmiths were able to heal the sick in many ways. This belief was connected to the air of mystery and supernatural origins that surrounded the trade. Some blacksmiths were known as blood charmers. This meant that the motion of their hands over an open wound would cause the bleeding to stop.

The number seven is often extremely significant in folklore. In fact, we have already noted that the blacksmith's role was one of the seven mechanical arts and one of the seven trades. When we examine healing within folklore, such as with witches and cunning men for instance, abilities were often said to be strongest in the seventh son of a seventh son. The same is true with the belief that a blacksmith could cure ills. This was most effective with a seventh-generation blacksmith.

> A seventh son has the power to cure the ringworm, and
> if there is a seventh blacksmith in a family he can do his
> choice thing.[9]

In some cases, more folk tales than actual concrete examples, the blacksmith succeeds where a doctor fails. At least two similar cases from Ireland have links with fairy folklore, as the patient is a sickly child thought to be a changeling (that is, a fairy from the fairy realm that they have secretly swapped for a healthy human child). The mother goes to see the blacksmith after having consulted with a doctor, who has not been able to provide a cure for the child. The smith advises the woman to go home and to declare that the woods next to their house are on fire.

The mother dutifully heads to her house and does just this, whereupon the changeling leaves the cradle, revealing itself to be a fairy rather than a child at all. Crying that its children will all be burned, the fairy rushes off and the mother's human child is returned later.

Sometimes the healing powers of the smith are attributed to the fact that he has a good relationship with the fairies. The receiving of skills or assistance from the fairies is not an uncommon trope in folklore. But of course, the fairy folk are often portrayed as mischievous and we may see this too in relation to blacksmiths:

> Deidle linkom doddie;
> We've gotten drucken Davie's wife,
> The smith o' Tullibody

This being cried by the fairies in glee as they carried the wife of the Blacksmith of Tullibody off up the chimney of their house.

In the same way that we may find 'white' witches who heal and 'black' witches who harm, where the blacksmith is able to cure he may also be seen to curse. When doing this, the anvil seems to play a significant role in the cursing process. It either had to be placed in a particular way, or turned in a certain fashion. One example from the National Folklore Collection of Ireland states that:

> If you want something to befall your neighbour, go to a blacksmith (and) get him to point the horn of the anvil to the east and to pronounce the curse.[10]

Another example from Ireland tells of a landlord who was found dead at the exact time of the turning of the blacksmith's anvil, his skin being all blackened. It was believed that he had been killed by the curse. In examples such as these, it is not unusual for little to be recorded about the mechanisms of the curse itself. This, of course, just adds more mystery to the magical reputation of the blacksmith.

IRON

Iron, the raw material with which the blacksmith fashions tools, weapons and other implements has long been believed to have supernatural powers. The ancient Egyptians thought that iron would ward off evil and, in many places, it is still used in this way. It is said that witches cannot cross iron, and that touching the metal will protect you from the evil eye.

The magical properties with which iron is said to be imbued are instrumental in the folk belief that it makes blacksmiths impervious to harm. William Yeats records this lovely story regarding a blacksmith who refuses to stop playing cards, which brings together the fairy folklore and the protective qualities of iron. The smith's wife employs the help of a boy to hide in the churchyard and try to frighten the smith on his way home:

> ... the boy did so, and began to groan and to try to frighten him when he came near. But it's well-known that nothing of that kind can do any harm to a blacksmith. So he went in and got hold of the boy, and told him he had a mind to choke him, and went his way.

But no sooner was the boy left alone than there came about him something in the shape of a dog, and then a great troop of cats. And they surrounded him, and he tried to get away home, but he had no power to go the way he wanted, but had to go with them. And at last they came to an old forth and a fairy bush, and he knelt down and made the sign of the cross and said a great many Our Fathers. And after a time they went into the fairy bush and left him.

And he was going away and a woman came out of the bush, and called to him three times to make him look back. And he saw it was a woman he knew before, that was dead, and so he knew she was among the fairies. And she said to him, 'It's well for you I was here, and worked hard for you, or you would have been brought in among them, and be like me.' So he got home.

And the blacksmith got home, too, and his wife was surprised to see he was no way (sic) frightened. But he said, 'You might know that there's nothing of the sort that could harm me.'

For a blacksmith is safe from all, and when he goes out in the night he keeps always in his pocket a small bit of iron, and they know him by that.[11]

Iron was sometimes used as a mechanism for healing, drawing on its protective qualities, and this was often in a charm form. To heal the sick, for example, a piece of iron would be placed on the afflicted area for a period of time before being taken off and nailed to a tree. This is very similar to a number of other charms for removing warts and the like. It was also said that if one bit on a piece of iron on the day before Easter, then one would avoid toothache for the rest of the year.

*St Dunstan Window, Leicester Cathedral. The right-hand window light shows
St Dunstan metalworking, and about to grab the devil with his red-hot tongs.
Source: Julian P. Guffogg (Creative commons licence)*

Similar ideas may be found in other belief systems too. One is undertaken by some Muslim mothers for a child who has hiccups. For this, any amount of money should be collected from seven people who should have the name Muhammad (again the magical qualities of the number seven in folklore are evident). This money is then taken to a blacksmith, who makes an amulet called a hazaazah. This is placed on the child's clothes to cure the hiccups. A charm such as this is quite controversial in the Muslim faith, and many would say that it should not be undertaken.

If you suffered from cramp, then this could be kept at bay by keeping a rusty sword by the bed. Presumably, even though it was rusty, the sword would need to be sharp, as another superstitious belief says that bringing dull iron into the house invites misfortune to come in after it. Similarly, a dull fire in the forge was also said to indicate poor trade. If you kept a hand-forged nail in your coin purse, however, then it was said that the purse would never be empty. I suppose this is true – after all, it would have a nail in it! Nails were also lucky when building. If you were constructing a barn or house, then the village blacksmith would hammer in the last nail such that it caught the rising sun. This would ensure that good luck was had every day.

Although it was often used for healing and considered to be good for protection, iron was also thought to have detrimental effects in some cases. Medicinal plants touched by iron are said to lose their healing properties. This is especially true of mistletoe. There also seems to be some connection between iron and food in folklore. When a member of the house passed away, all of the food items in the house should be pricked with a piece of iron in order to stop death from spoiling them. Iron was also used to help butter to form. This would be done by placing a red-hot iron rod or a poker into the churn.

One Irish story tells of a blacksmith employing a cure in order to find a person who was stealing butter in a town. The blacksmith made a horseshoe and a set of nails. These two items had to be made by heating the iron used to different temperatures. They were then put underneath the butter churn. The story then takes a twist rather reminiscent of witchcraft tales, as the person responsible for taking the butter is discovered in the form of a hare. This is a common motif in folklore where a witch undertakes some act and flees in the guise of a hare. Sometimes the hare is injured and the witch is discovered in human form the following day, exhibiting the same injury. In this case, however, the story ends with everyone in the town getting their butter back.

An interesting side note here is that this is not the usual way in which stories of stealing butter are resolved in folklore. More common is the criminal being harmed by someone heating a piece of iron and putting it into the milk. Note the similarity to the method of helping butter to form mentioned above.

John McNab Currier, a nineteenth-century physician, was also a collector of folklore. He recorded a story in the *Journal of American Folklore* in 1891 about a blacksmith who was a believer in witchcraft. The events were said to take place in a town in New Hampshire sometime between 1845 and 1855. The blacksmith uses the protective power of iron in the form of a horseshoe to prove his suspicions about a visitor are correct:

One day a man came into [the blacksmith's] shop to get a small job of work done forthwith, being in a hurry to return to his work. The blacksmith … nailed a horseshoe over the door, believing that if so possessed [ie if the man was a witch] he would be unable to pass out of the shop under it. The man's job was immediately finished; but, instead of starting for home, he lingered in the shop nearly all the forenoon, and seemed in no hurry to get away, pretending that he was waiting to see a man who, he thought, would shortly pass that way. This sudden change in the plans confirmed the blacksmith in his suspicions of the man's character, and he removed the shoe from over the door, and the man started for home at once.[12]

Where iron is normally employed to keep witches out of a property, here the blacksmith does the opposite and traps one inside. In a similar way, cemeteries were often surrounded with an iron fence because it was thought that this would contain the souls of the dead and stop them from wandering.

Although there are many connections between blacksmiths and the devil, witches or the supernatural, we do find connections of a more religious type as well. In many places it is considered to be wrong for a blacksmith to handle shoes and nails on Good Friday. Nails would not be hammered on that day out of respect, in remembrance of the crucifixion. Similarly, it is said that the blacksmith is lucky, as is his place of work, because he refused to make the crucifixion nails. The tinker (a wandering tin-smith) is always down on his luck because he was willing to do so.

FOLK MEDICINE

We have already seen that water in the forge is seen to be curative. It is also said that the blacksmith would be able to regain strength and stamina by washing his hands in the forge water. This was considered to be the case because the Virgin Mary had once blessed a blacksmith's water.

In both Ireland and in some parts of Britain, a charm called the 'Horseman's Word' was said to give blacksmiths (and other horsemen) absolute control over horses. Many people believed that those who worked with horses closely did so in magical ways, such as by whispering a secret word into the ear of the animal. Semi-secret societies, not unlike the freemasons, were established for these people and were especially prevalent in Scotland. These also became known as 'The Horseman's Word'.

Blacksmiths would also spit on their hands in order to help to subdue an unruly horse. This custom may be found recorded in the poem 'Blacksmith' by Thomas Miller:[13]

As when a smith and his man, lame Vulcan's fellows,
Called from the anvil or the puffing bellows,
To clap a well-wrought shoe, for more than pay,
Upon a stubborn nag of Galloway,
Or unbacked jennet, or a Flander's mare,
That at the forge stands snuffing up the air,
The swarthy smith spits in his buck-horn fist,
And bids his man bring out the fivefold twist,
His shackles, shacklocks, hampers, gyves, and chains –

When if a carrier's jade be brought unto him,
His man can hold his foot while he can shoe him.

Most by-products from the smithing processes could be put to good use. We have already covered water and horseshoes, but even the black oxide known as anvil dust that contributes to the blacksmith's name had multifarious uses. In a letter from the Secretary of War to the Government of the United States Army, released on 26 December 1820, it was noted that anvil dust should be sifted well through an old stocking and moistened with sweet oil or emery as an effective method of removing rust from firearms. Another practical use, recorded three years earlier in the *Gentleman's Magazine* of 1817, was to sprinkle it onto the tar when coating a paper roof. Gardeners could mix equal parts of anvil dust and lime with a larger quantity of horse manure or black earth from the woods in order to create a treatment for 'rose bugs', or aphids.

Anvil dust is an important ingredient within folk magic undertaken by practitioners of African-American Hoodoo. They believe it to be a lucky substance and it is often carried upon the person, sometimes in a red flannel bag. Some people sprinkle it on a pair of lodestones in order to ensure that luck stays with them. It is also sometimes used as an ingredient in 'Goofer Dust' mixture. This is a traditional Hoodoo hexing material. In rarer cases, anvil dust may also be ingested, although this is not to be recommended. It would be mixed with molasses, for example, as a tonic. This was presumably seen as a way to deal with someone who appeared to have an iron deficiency.

Anvils incorporate a deep hole, called a hardie hole, which is used to hold specialist forming tools, or sometimes as part of the punching process. One old method of testing the anvil before use to make sure that it had no defects was to ignite gunpowder on top of it. If the anvil did not crack then it was good to use. Similarly, an old tradition on St Clements Day was to fill the hardie hole with gunpowder before driving a wooden plug in on top of the powder. A hole would then be bored through the wood, a little more powder poured in and the top ignited. This was known as 'firing the anvil'. In more modern practices, anvil firing looks to launch the entire anvil into the air. It is not clear whether the older 'firing' of the anvil did this, or just caused a loud report. Nowadays, two anvils are used, with one being placed upside down on top of the other as a projectile.

Moving on from the anvil and its dust, we may find that even the cinders from the fire may be employed in other ways. One old custom told of dropping red-hot coals from the fire into the water, which is then to be used to wash a new-born child. We must presume that they allowed it to cool before they used it!

Cinder-tea (which is not actual tea but refers to water that has been heated using a hot cinder) was thought to be effective for curing both colic and flatulence in children. The *Folklore* journal for 1901 records this story from Wakefield, told by a Mrs W.M.E. Fowler:

When one of my brothers was a few weeks old – in 1867 – he was suddenly seized by an attack of convulsions during my parents' temporary absence from home. On my mother's return, the nurse rushed to meet her and told her of the baby's illness, continuing, however, to say that the infant was now sleeping quietly, and would certainly recover, as she had done all that was necessary or possible in such a case – namely, had dropped a hot cinder into a cup of water, with which she had baptised the child by making a cross on his forehead, afterwards giving him the remainder to drink; thus, apparently, in her own eyes bringing both religion and science to bear on the case. The nurse was, I believe, an Irishwoman, who had been brought up in Yorkshire, where she would have many opportunities of learning the well-known healing power of 'cinder-tea'.[14]

The blacksmith's apron was an important part of his apparel, and has some interesting folklore attached to it, too. We have already seen the importance that the smith used to have in a community. He would receive early invitations to all of the local social events, be given grain and fruit at harvest time and good meat when animals were killed. But another mark of his status was that, when wearing his apron, the smith had full permission to shake hands with a king. He could also freely cross any private property while so attired. There are various pieces of lore said to explain the slits that are cut in the blacksmith's apron. One, coming from the story of the king choosing the tailor over the blacksmith as the most important trade, which was related earlier, says that the tailor crawled under the table during the feast and cut a fringe in the apron with his new shears. Another says that they signify a lion's paw; if a blacksmith had lions displayed on his smithy then that constituted a freehold on the property.

A third is very similar to the story of the blacksmith and the king, but with more biblical connection. It comes from the building of King Solomon's temple, and it says that after the construction was completed a supper was held for all of the tradespeople, but the blacksmiths were not invited. As a result of this, they stopped work in protest. In time of course, as in the story with St Clement, all of the tradespeople needed tools making or repairing. There was nobody to do this and no further work could be done. King Solomon therefore intervened and put on a second supper. At this, he had the fringe cut into the apron, and it was gilded as a mark of respect to the smiths. Another piece of Irish clothing folklore states that a woman who was in labour would make sure that the delivery of the baby went ahead without any problems by wearing a smith's waistcoat.

In more modern times, many of these old customs have all but died away – largely as a result of more mechanised manufacturing processes and changes in technology, which have led to the gradual demise of the blacksmith. The craft is still carried out by smiths and by farriers, but their numbers are fewer and many have diversified into other areas, such as the production of ornamental ironwork. There is no longer a blacksmith at the centre of many communities. And despite the belief that it was unlucky to build a house on the site of a forge, if you take a walk through many small villages now you will find a property named 'The Old Forge' or similar.

The figure of the blacksmith may linger occasionally as a folk memory. If you see one in your dreams, then this is supposed to signify that you will be successful in your chosen career. If you actually are the smith yourself in the dream, then this is said to mean that you have a desire to learn new skills or discover new ideas. But whereas the American Indians looked upon blacksmiths as being allied to the spirits, and referred to them as the ghosts of iron, to us now they are more becoming ghosts of a past time, when life was simpler and those with such skills were to be revered.

As we began with a poem, so it is probably appropriate to close with these lines:

The anvil's roar
We'll hear no more;
The forge is silent now[15]

Chapter Three Sources

1. Sieroshevsky, V.L., 'The Yakuts', 1896, St Petersburg
2. Popov, A., 'Consecration Ritual for a Blacksmith Novice among the Yakuts', The Journal of American Folklore, Vol 46, No 181, American Folklore Society
3. www.english-heritage.org.uk/visit/places/waylands-smithy/history Accessed 4 September 2019
4. Graça da Silva, S and Tehrani, J.J., 'Comparative phylogenetic analyses uncover the ancient roots of Indo-European folktales' 2016, Royal Society Open Science
5. Wintemberg, W.J., 'German Folk-Tales Collected in Canada', The Journal of American Folklore, Vol 19, No 74 (Jul – Sep 1906), American Folklore Society
6. Sawyer, F.E., '"Old Clem" Celebrations and Blacksmiths' Lore', The Folk-Lore Journal, Vol 2 No 11 (Nov 1884), Taylor & Francis Ltd on behalf of Folklore Enterprises
7. ibid
8. 'Legend of the Ringing Anvil or Why We Ring Our Anvil and, When Comes The Luck of the Horseshoe', www.anvilfire.com/21centbs/stories/ Accessed 17 August 2019
9. Yeats, W., 'Writings on Irish Folklore, Legend and Myth', 1993, Penguin Classics
10. 'Blacksmiths and the Supernatural' irishfolklore.wordpress.com/2017/03/13/blacksmiths-and-the-supernatural Accessed 17 August 2019
11. Yeats, W., 'Writings on Irish Folklore, Legend and Myth', 1993, Penguin Classics
12. McNab Currier, J., 'Contributions to New England Folk-Lore', The Journal of American Folklore, Vol 4 No 14 (Jul – Sep 1891), American Folklore Society
13. Miller, T., 'Rural Sketches', 1839, available at https://archive.org/details/rural-sketchesoomillgoog/page/n13
14. Peacock, M., Carson, K. and Burne, C.S., 'Customs Relating to Iron', Folklore, Vol 12 No 4 (Dec 1901), Taylor & Francis Ltd on behalf of Folklore Enterprises, Ltd
15. 'The Village Blacksmith', garrafrauns.com/stories-folklore/village-blacksmith Accessed 17 August 2019

4

BEER AND BREWING

There was three kings into the east,
three kings both great and high,
and they hae sworn a solemn oath
John Barleycorn must die.
They took a plough and plough'd him down,
put clots upon his head,
and they hae sworn a solemn oath
John Barleycorn was dead.
But the cheerful Spring came kindly on'
and show'rs began to fall.
John Barleycorn got up again,
and sore surprised them all.

Robert Burns' poem, of which those lines are the beginning, was written in 1782. The named character, John Barleycorn, is a metaphor for the barley crop, which is harvested every autumn and is an integral part of the brewing process. The unpleasant events that John Barleycorn suffers are representative of the constantly turning cycle of planting, growing, harvest and death – leading to rebirth in the following spring. These important ideas of death and rebirth suggest a much earlier source for the John Barleycorn metaphor, in line with the older nature religions and their cyclical calendars. The earliest written example does indeed come from a document called the Bannatyne Manuscript in 1568. But it is possible to speculate at a much earlier source than this.

There is a figure in early Anglo-Saxon religion called Beowa who would seem to be a parallel with the character of John Barleycorn. He is associated with agriculture and particularly with the threshing of grain. Beowa is also the Old English word for barley. Aside from water, barley is usually the main ingredient used in the brewing process. It is used because ancient brewers found that the grain supplied its own enzymes when it was germinating. This replaced the methods employed by Inca women, for example, where grains were chewed as part of the preparation process. The natural enzymes in the saliva broke down the starch as required.

We can see then, that the brewing of beer undoubtedly stretches back a long way into our history. And as we have already learned in the previous chapters, where a trade or profession has early roots and is still important today, it naturally becomes culturally significant and develops much folklore and tradition alongside it on the way.

MYTHOLOGY AND HISTORY

We can find evidence of brewing beer and other alcoholic drinks in Ancient Egypt. The Greeks noted that beer as a drink came out of Egypt first. The development of the drink probably came about because of the unsanitary nature of the water in the area, which was unsafe to drink because of the lack of sewer systems to remove waste. Although wine was also drunk, beer was popular in many areas where grapes could not be cultivated.

A barley-based beer was the normal drink at this time, but other types were brewed for special occasions. Barley and other grains became so important in Egypt that they developed into a form of currency. The brewery owned by Ramses, for example, gave 10,000 barrels of beer each year to the administrators of the temple. The drink obviously gave much pleasure to many, as one Sumerian poet wrote in approximately 3,000 BC:

In the ... reed buckets there is sweet beer,
I will make cupbearers, boys, [and] brewers stand by,
While I circle around the abundance of beer,
While I feel wonderful, I feel wonderful,
Drinking beer, in a blissful mood,
Drinking liquor, feeling exhilarated,
With joy in the heart [and] a happy liver –
While my heart full of joy,
[And] [my] happy liver I cover with a garment fit for a queen![1]

Both here and in Mesopotamia, beer also connected with healing and with religion, showing its significance. Beer was used in Egypt to strengthen gums, put onto wounds and even administered as an enema.

The list of deities who have a connection to beer and brewing is a very long one, running to at least one hundred entries across all of the world's cultures.[2] We can therefore only look in more detail at a few examples from mythology. Religious festivals were a time when beer consumption would increase, because it was generally believed that getting drunk would bring you closer to the gods. In many cases, beer was seen to have been a gift from God, or from the gods, depending on the culture (see below). One medieval term for yeast, in fact, was godisgoode.[3]

Staying for a moment with the Egyptian culture, most people would normally associate the god Osiris with his role as Lord of the Underworld. In this position he would oversee death, the afterlife and resurrection and here we can see a similarity with this cycle of birth, growth, death and rebirth relating to the crops, which was mentioned above. So it is interesting to note that Osiris was also the god of beer and wine. Along with his wife Isis, who is also seen as mistress of these beverages, Osiris would be linked with the agriculture in the Nile Delta. It was said that he was responsible for imparting the knowledge of the brewing process to the Egyptians, and so, when they died, people would often be buried with beer as an offering to the god.

The goddess of beer (and therefore of a higher rank than Isis as mistress) was Nephthys. This was because she was thought to be the source of rain, and so also intrinsically linked with the Nile river. Water being the main component of beer, she was an important figure, but she did not oversee the brewing process. This task fell to Tenenet. Women made both bread and beer in Egypt, two vital elements of the diet at that time. And so we find that Tenenet, whose name may possibly come from the Egyptian word for beer itself, tenemu, was goddess of both brewing and childbirth.

For the Zulus of South Africa, another goddess created beer. This was Mbaba Mwana Waresa and, in a similar way to the Egyptian mythology, she was also said to have given the knowledge of the brewing process to those who took up the trade. There is a traditional folk tale that tells how this came about.

Wooden tankard retrieved from the Mary Rose. Source: Mary Rose Trust. (Creative commons licence)

Mbaba Mwana Waresa searched the heavens for a husband, but could not find anyone who she thought was suitable. So, she travelled to the mortal world to see whether there was a South African man who would be a better partner for her. A herdsman called Thandiwe saw Mbaba Mwana Waresa and was enamoured of her, so he tried to win her attention by serenading her with a lovely song. This worked, but the goddess was still concerned that Thandiwe would not be the right man to marry. She therefore conceived a test for the herdsman.

Mbaba Mwana Waresa sent a beautiful woman to Thandiwe, whilst she herself changed shape and came to him in the guise of an old woman. Despite the trick, Thandiwe recognised the goddess immediately and rejected the younger woman in favour of her. By this, Mbaba Mwana Waresa knew that she had chosen the right partner and they were soon wed.

The other gods were angry that Mbaba Mwana Waresa had forsaken their realm and married a mortal, on whom they looked with some derision. In order to overcome this problem, the goddess created beer as a way to make humans feel closer to the gods (again, much like the Egyptian culture). She shared her brewing knowledge with the mortals so that they could continue to make the drink in her absence and remain on good terms with the immortal realm.

In traditional Zulu practices even today, the woman who brewed the beer will remove the froth from the top and pour it onto the ground as an offering to both the spirits and the ancestors. Men should remove their hats when drinking, as a mark of respect to the goddess who first gave them the knowledge of how to brew.

In Norse traditions Aegir, the God of the Sea, was also the brewer to the other Gods of Asgard. It may be that his association with brewing comes about because the white foam on ocean waves bears some resemblance to the top of a mug of beer.

Aegir's feast hall was at the bottom of the sea. He was said to serve the best ale in all of the Nine Worlds. Not only that, but it was served in drinking horns that never ran dry; refilling themselves before they were emptied. This naturally made the Hall a favourite visiting place for the Gods. It was the only location where they could not follow their normal favourite pastime of fighting each other. This was because they could not risk the punishment for doing so – eternal banishment from the Hall – and therefore lose the endless beer vessels.

Greek and Roman gods always figure largely in any mythology. Beer and brewing is no exception, with the Greek goddess Demeter being the barley-mother. This links her also with the agriculture and the harvest and so, once again, we return to the start of this chapter and the tropes of death and rebirth in nature. In the Roman pantheon, this same mother goddess figure was named Ceres and from her the Spanish word for beer, cerveza, emerges.

Amongst all of the myths and legends associated with beer, one of the more significant is probably that of Gambrinus, the King of Beer. Folk tales about Gambrinus can be found across Western Europe, and he has become the patron saint of beer and brewing. Both drinking establishments and the beers served in them have been named after him.

Gambrinus, a handsome apprentice to a glass blower, is in love with Flandrine, the daughter of his master. But she will not return his affections before he becomes a man with status.

Gambrinus becomes very unhappy and so he leaves town in an effort to try and forget about his unrequited love. He travels around playing his violin, in which he is very proficient. When they hear about his success, the folk of his town beg Gambrinus to come back and give a concert. He agrees to do so but partway through the recital, which everyone is in awe of, he spots Flandrine in the audience and can no longer concentrate on his playing. The music becomes dreadful to listen to and the audience become very angry and begin to riot. Gambrinus is thrown into jail, where he decides to hang himself as he is so miserable. Before he can do so, however, the devil appears in Gambrinus's cell, disguised as a hunter, and offers him happiness in return for his soul, to be collected thirty years from that point. Gambrinus asks for Flandrine's love but the devil cannot meddle in affairs of the heart. Instead, he offers to give Gambrinus something to take his mind off the girl. The contract is duly signed.

The devil has given Gambrinus skills in gambling, and he becomes very rich through doing so. He tries to approach Flandrine once again, the effort to forget her having been wholly unsuccessful, but she still will not have him. This time she says that it doesn't matter that he is rich, he is still a nobody. She tells him to return when he is a King or a Duke.

Gambrinus leaves the area once again, and for a second time the devil comes to him. This time the devil shows Gambrinus how to build a brewery and make beer. Tasting the results at some length, Gambrinus is happy again. The devil sends him on his way, after having given him an enchanted chime that plays such amazing music that anyone who hears it is compelled to dance to it.

Once again, Gambrinus returns home. He sets up the brewery and invites the townspeople to try it, but they find it bitter and unpleasant and mock him for producing it. At this, Gambrinus plays his chime and forces everyone to dance to the music for several hours until they become very thirsty. This time, drinking the beer to quench their thirst, they grow to like the taste and the brewery becomes extremely popular. Because of the success, Gambrinus becomes both rich and titled, which is everything that Flandrine wanted of him. She comes to visit him, but now he has sadly forgotten her and she just gets sent away with her own beer.

Gambrinus continues to be rich and successful, and not altogether sober, for the next thirty years until finally the devil arrives to collect his soul. Gambrinus plays his chime to welcome the devil and, fortunately for him, the devil also falls under its magic and is compelled to dance to it. He is kept dancing for hours and, despite his pleas, Gambrinus will not stop playing. Eventually, the devil offers to tear up the contract in return for being freed from the spell and returns to hell. Gambrinus gives the devil a barrel of his own beer as a parting gift.

These stories of being given skills by the devil are not uncommon in folklore. In the last chapter we learned how the blacksmith was said to have acquired his abilities and this story is not too dissimilar in many ways. In the case of the blacksmith, the story of trapping the devil related to a real figure, St Dunstan. Whether Gambrinus was based on a real person is up for debate, although there are some good candidates. John I was Duke of Brabant (the most common title conferred on Gambrinus in versions of the tale) and was also known for liking beer. His Dutch name, Jan Primus, sounds convincingly similar when spoken out loud.

Other possible figures are John the Fearless, who was Duke of Burgundy, and an obscure Germanic King called Gambrivius, who was said to have been taught to brew by the gods. This latter suggestion has been discounted by many historians and, in light of the fact that the story is quite widespread across Europe, Gambrinus could have drawn from more than one person for its origin story.

The story of Gambrinus is well known in areas of Belgium. This country is noted for the traditional brewing of beer; so much so that UNESCO recognised its cultural significance in 2016 and added it to the list of Intangible Cultural Heritages.

The knowledge of how to produce the range of craft beers found in Belgium has been passed down through generations. The consumption of beer is such a large part of life that children used to be offered a very low alcohol-beer known as 'table beer' as one of their school lunch drink choices. This was stopped in the 1960s.

BREWING IN THE MIDDLE AGES

In the early period in England, rather like in older times in Egypt, much brewing was undertaken by women. Although originally not particularly well controlled, the strength of the beer being produced was regulated by statute in 1267. If the brewer did not comply with the terms of the statute, then they could be fined under one of three offences. These were: the selling of bad beer, the selling of too small beer or selling beer either in unmarked measures or at a price that was different to that set by the assize. The same statute also set rules governing bread production (see Chapter Five). Tasters were in the employ of the local Lord to check on beer and bread production and, in fact, there is still a traditional ceremony in Devon harking back to these times: the annual Ashburton Bread Weighing and Ale Tasting.

Walter de Stapledon was Bishop of Exeter in 1310 and, at the same time, also acted as Lord Mayor of Ashburton, which put him in a good position to be able to obtain a Royal Charter from then-king, Edward II. The Charter granted the people of Ashburton the right to hold a market on a Saturday as well as an annual three-day fair. Ashburton Fair originally took place on 'the vigil, the day and the morrow of the Feast of St Lawrence' – 10 August being the Feast Day. Three years later, in 1313, the Feast Day was changed to that of St Martin, placing the Fair around 12 November.

Being a stannary town, Ashburton has two ancient courts whose roles are now largely ceremonial. These are the Court Leet and the Court Baron. The latter is made up of tenants of the town and elected such positions as the Viewers of the Water Courses, Viewers of the Market, Inspector of Trees, Pig Drovers and Scavengers. The Court Leet is formed of freeholders and elects two very important positions: Ale Tasters and Bread Weighers. Concerns over ale quantity were already in place when the charter was granted. Mentions of this concern in the Magna Carta had led to Henry III introducing the Assize of Bread and Ale in 1267. Records show that an Ashburton brewer was indeed fined a penny in that century for selling 'bad ale'.

Yard of Ale. Source: thefreefood.net (Creative commons licence)

A whole ceremony formed around ale tasting and bread weighing, with various checks being made. Ale would be poured onto a wooden bench on which the Ale Taster would sit, wearing leather breeches. If the clothes stuck to the bench, then it was a sign that the ale had been mixed with sugar. Of course, taste and smell were also important, and today arguably form a larger part of the ceremony. Originally part of the town carnival, the Ceremony of Ale Tasting and Bread Weighing moved away to become its own entity in 1986 and now takes place on the third Saturday in July.

The price of beer would have traditionally been linked to its perceived quality. At the 1405 Feast of the Holy Cross Guild in Stratford, for example, the prices were 'gude beere 1½d, peneyale 1d, small ale 1½d a gallon'. But old folk rhymes record actual names for some of the beers brewed. Some of these include:

In Ilmington:

> Black strap,
> Ruffle-me-cap,
> Fine and clear,
> Table beer.

In Whitchurch, two bushels of malt produced:

> Forty gallons of clink-me-clear,
> Forty gallons of table beer,
> Forty gallons of Rat-me-Tat,
> Forty gallons worse than that.[4]

And a mile away in Alderminster, a woman named Keyte brewed:

> Double ale, single ale, very good ale,
> Twine-in-the-belly,
> Twice as many,
> Tip tap, worse than that.
> Pin.

One tale tells that she once gave a man 'Tip Tap' to drink, and after tasting the beer he began laughing. On being questioned by the brewer as to what was so funny, he replied that he could not work out how she managed to brew two more beers that were worse than that one.

Brewing at this time tended to take place mostly in the months of March and October. As well as the normal-strength beers, a much weaker drink called tilly willy was brewed at both of these times for the children to drink. The October brewing produced plain beer, but in March other ingredients would sometimes be added. One March tilly willy included ale hoof, Herif and common nettle shoots (as well as blackcurrant leaves if the plants had grown enough) and this ale was said to prevent 'spring rash'.[5]

The monks of the Abbey of St Peter in Oudenberg, Belgium, were well known for brewing beer in the eleventh century. Beer was an important part of medieval life both here and elsewhere. As with the Egyptians much earlier, beer was a substitute for drinking water, which was generally unfit for consumption because of contamination in the streams. Estimates put the daily intake of the weaker 'small beer' at somewhere around 1.5 litres per person. Most brewing at this time was done by families rather than in any commercial capacity. One superstition connected with this practice suggested that if you were making small beer and you thought of someone who had been in a very bad mood recently, then the resulting beer would turn out to be very mild.

It was really the abbeys who brought about the first of the larger breweries. Although it was generally believed in the Middle Ages that beer was not particularly good for you, the founder of the Abbey at Oudenberg, Bishop Arnold, was said to have told the local peasants to drink it instead of water in order to take advantage of its 'gift of health'. Although we may view this to some extent as a cynical marketing ploy on behalf of the abbey, it certainly was the case that switching from water to beer during periods of widespread illness did help to save more lives. So, just how efficacious was it thought to be to drink beer?

BEER AND HEALTH

Natural ingredients have always been used for the treatment of ailments, and since modern processes of synthesis and other chemical processes had not yet been developed, they were obviously the sole method of pain relief and treatment by medicine. The simple fact that the processes that go into the brewing of beer make the end product safer to drink than the contaminated water means that there are immediately obvious health benefits. From this starting point, brewers learned as they developed their skills that beer was not only a liquid that might have some beneficial health properties, but that it was a means by which other medicinal items could be delivered. Healers who worked with plants, berries and other natural ingredients were able to provide the knowledge as to which items were effective as particular cures. For their part, the brewers knew that some of these could be dissolved in beer, where they might not have dissolved in water alone. Patrick McGovern noted a number of examples from ancient Egyptian medical documents where beer was used in this way.[6] Spices such as coriander and cumin, herbs such as dill, and plant derivatives including aloe and mandrake all appear as additives in this way. Chinese medicine has plenty of examples of the same process. Greek physicians followed the same ideas and among their texts we can find such recipes as:

> For good and plentiful breast milk: mix unripe sesame plant fruit or crushed earthworms and palm dates into the beer.

> For poisonous asp bites: use beer with crushed garlic as an emetic to induce vomiting.

> To expel intestinal worms: use beer to soak a herbal suppository.

> To treat coughs: beer may be drunk warm with salt.[7]

The medieval period is rich in unusual medical treatments, and beer features in many of these, including hot ale (chest pains), new ale (insomnia), old ale (lung disease) and Welsh ale, with other ingredients for a number of conditions. It was even used as a scalp treatment in an attempt to kill off head lice.

Beer continued to be associated with health long after the Middle Ages. Two correspondents to the *Sunday Times* in 1956 noted more recent examples from their own families:

My father was at Christ's Hospital in London from 1861–69. For breakfast the boys had beer, and scrug (bread) which they were supposed to dip in the beer. As the big boys took the smaller boys' beer and scrug, they often went without breakfast. (Mrs) D.M. Mabey, Purley.

In her younger days, my great-grandmother, born 1815, a daughter of Dr Henry Hilton of Liverpool, drank beer for breakfast. She attributed her fine complexion to the fact, and believed that English-women's complexions began to deteriorate when they substituted tea. J.N. Deacon, Stanmore Common.[8]

For many hundreds of years, advice has been given to breast-feeding women regarding alcohol intake. Like the advice itself, the folklore surrounding this is often contradictory. Some say that drinking whilst breastfeeding can increase the risks of the child suffering from 'acquired alcoholism' later in life. Others contend that drinking beer when lactating increases a mother's milk supply and helps the baby to get stronger. Brewers actually produced and marketed a tonic in the early part of the twentieth century to do just this. It has not been proven that this is the case scientifically, but it is true that drinking beer increases a hormone called serum prolactin, which is necessary for the production of milk.

Before the inclusion of hops in the brewing process, plants were also used to help prevent bacterial infections in the beer, which would cause it to taste incredibly sour. Among these were rosemary and bog myrtle. Many of these sorts of traditional plant remedies have died away as modern medicine has developed alternatives, but there is enough circumstantial evidence associated with some of the beer remedies to encourage modern scientists to investigate further. One programme, for example, is looking at traces of beer from grave goods to search for possible properties to treat cancer, and has had some encouraging results.

Where there are benefits in ancient beers, there are similar possibilities with modern ones, and chemists are also looking at whether some substances that come from hops are beneficial in treating diabetes. Beer is also seen as a good preventative for some conditions. Cardiovascular disease and kidney stone risks may be reduced by the drinking of beer. Intake, of course, does need to be in moderation for these effects to be positive.

HANGOVERS

Should your alcohol consumption be less than restrained, you should not be surprised if you suffer the after-effects the following day. Whilst the brewing and drinking of beer go back to Ancient Egypt or earlier, as do the problems of over-indulgence and the aids suggested to relieve them, the common term for the morning after is surprisingly recent in the English language.

'Hangover' as a term for the effects of excessive drinking is traced as first entering the lexicon in 1902, with the slightly earlier meaning of 'a thing left over from before' on which it is based being traced to 1894.[9] This in itself is curious, considering the Anglo-Saxons would have suffered from the problem many years before this. English seems to be alone in taking a long time to name the condition suitably. Most other languages have very old terms for the hangover. Bufo Yamamuro, an ex-director of the Christian Temperance League in Tokyo, noted for example that the Japanese term *Futsuka Yoi* (translating as 'second day drunk') is approximately 2,000 years old.[10]

There are many variations in the ways that people believe they can treat their hangover; probably as many as people who have suffered from them in the first place. One of the most well known is the idea that having another drink of what you were enjoying the previous night will ease the headaches. Commonly known as 'the hair of the dog', the phrase goes back a remarkably long way into history. This is because the idea comes from applications of sympathetic magic – influencing a person or condition using an item associated with it. Pliny, writing in the first century AD, said that there was a custom of curing a bite from a dog by burning some hairs taken from the tail of the dog that did the biting, and then rubbing the ashes into the wound.[11] This is possibly the original reference from which the idea of treating the hangover comes. Ozark superstition, even in the twentieth century, said that you should consume some of the hair from the dog that bit you, so we can see a link to ingestion in this case at least.

Folklore has accumulated so many hangover remedies over time that they can be broken down into sections and listed very easily. Food and drink are quite common as treatments, and in many cases these seem to link with old folk medicine cures for other afflictions. Onions, for instance, were used to treat nervous conditions but also appear in culinary treatments for a hangover, such as:

Slice a big raw onion and drown it in vinegar and olive oil.

Eat onion soup instead of breakfast, with grated cheese and lots of pepper.

Eat a white Bermuda onion like an apple. Wait half an hour, then take a good shot. But don't drink before a half an hour or you'll get sick.

Eat a good rare steak with raw onions.[12]

Steak appears many times: raw steak, rare steak and steak sandwich are all mentioned. This is probably due to the vitamins and iron in the red meat. Tomatoes and eggs are also frequently mentioned. Tomato juice is very well known as a remedy for a hangover, and this may be connected to an older folk belief that it will also sober you up whilst you are still drunk.

It is interesting that beer itself is very often described as being effective for hangovers. This is firmly in the 'hair of the dog' arena. Beer and tomato juice together are often cited as being drunk in this way. Bar folklore says that warm beer should be drunk the morning after, and therefore anyone coming into a pub and ordering a warm beer rather than one from the fridge must already have a hangover.

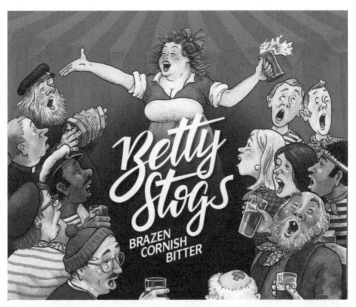

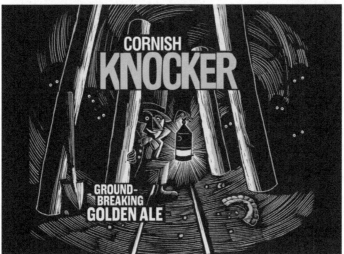

Original artwork for Betty Stogs and Cornish Knocker ales. The names were taken from folkloric characters. Source: Skinners Brewery

Whilst there are no effective pharmaceutical cures for the hangover, many people believe that aspirin or baking soda will ease the symptoms. Others think that certain vitamins are good. But the most effective suggestion is probably that of Don Brown, a sales manager from Detroit, who when asked for his remedy as part of a survey collecting folkloric hangover cures, said:

> I'm 63 years old come Sunday, and I've had my share of hangovers, but if you want to know the truth, there's no cure except time.[13]

He did go on, however, to suggest that eating raw cabbage helped a little too!

How many of these hangover cures actually work and how many are superstitions is probably down to the individual who is trying them. Even if the placebo effect is responsible in some cases, it will help to add to the ongoing belief in the treatments. But hangovers are not the only element of drinking, or indeed in the production of beer, over which superstitions have arisen.

SUPERSTITIONS

The Greeks believed that you could receive a warning that danger was going to come to a loved one who was distant from you. If the beer in the mug that you were holding started to bubble, then that would be the signal that something had taken place.[14]

It was said that when brewing, you could stop the beer from turning sour by putting a coal into the liquid before adding the yeast. Similarly, folklorist J. Harvey Bloom noted, in the 1920s, that pieces of iron were placed on the tops of casks of beer that had recently been brewed in order to protect them from being soured during a thunderstorm.[15] If you wanted to know if beer was particularly good, then one sign was the froth sticking to the inside of the glass.

In 1620, Yorkshire author Sir John Melton wrote a satirical play called 'The Astrolagaster, or the Figure-Caster'. The piece was a swipe at the astrologers and fortune-tellers, who he considered to be charlatans. In this text he says that, 'if the beer fall next a man it is a sign of good luck'.[16] The question that we must ask here is whether this was a superstition that existed prior to the writing of this play, or whether Melton included it as a fictional piece of belief. In terms of folklore, it is not a matter of too much concern; either way the superstition would continue to proliferate after this date.

A Bohemian superstition says that to upset beer means that there will soon be a christening in the house. Another more general old superstition suggests that one should drink beer in a shop in order to ensure that it will be prosperous.

In Czechia, it is said to be bad luck to pour a beer into a glass that previously contained a different kind of beer. If anyone was qualified to know this, then it should be the residents of the Czech Republic. According to research undertaken by a Japanese drinks company, the country has been the world's number one beer consumer per capita for the last twenty-four years.[17]

In Russia, a fascinating story explains why it is said to be bad luck to put empty beer bottles on the table. In 1814, says the story, Russian Cossack soldiers were pushing Napoleon's armies back through France. Whilst in that country, the soldiers noticed a way that they could get free beer in the restaurants. They worked out that the Parisians calculated how much was owed by counting up the empty bottles on the table rather than recording how many bottles were ordered. To counter this, the Cossacks kept some of their bottles on the floor when they had finished with them, rather than putting them all on the table. When they returned to Russia, it is said that they brought the custom back with them, where it became the superstition about good luck. It is also traditional in Russia that the last person to leave a party must have a final drink. This is known as *na pososhok*, which literally translates as 'on a small walking staff'. The sentiment behind this is to wish for a safe journey home by blessing a walking stick, which traditionally many people used.

Na pososhok is sometimes used as a toast at the end of a party also, for the same reason of wishing the guests safe travels home. Toasts are often thought of as being generally done with wine or spirit-based drinks, but they may equally be undertaken with beer. What is generally considered to be bad luck is to perform a toast using only water. One example of this superstition comes from the *Mess Night Manual*, published by the US Navy:

> Although civilian practice is more permissive, in the military, toasts are never drunk with liqueurs, soft drinks, or water. Tradition is that the object of a toast with water will die by drowning.[18]

In Spain, there is an altogether different curse associated with making a toast using water. Here the tradition is that the person doing so will suffer seven years of bad sex. This surprising detriment appears in the folklore of many other countries too, in association with toasting. Both the French and the Germans hold that the same fate will befall those who break eye contact during a toast. Back in the Czech Republic, you should never cross arms with anyone or the same affliction will result.

In Hungary, toasts are not made with beer at all. The reason for this is again based in war, but this time goes back to the Hungarian Revolution of 1848. It is said that because the Austrians raised glasses of beer to cheer the repelling of the Revolution, Hungary pledged that they would not use beer in this way for 150 years. Although technically it should by now be acceptable to do so again, over the long period of time it has become traditional that it is not done, and it has therefore now entered into superstition that it would be back luck to do so.

DRINKING CUSTOMS

Customs in general reflect 'our values, our beliefs and the trends in our societies'[19] and this applies equally to those surrounding the drinking of beer. This is hardly surprising, considering the long history of its production and consumption, that has already been discussed. Egyptian royalty would drink their beer through thin gold straws, which were designed to ensure that they did not drink the dregs of the beer. In African villages, straws are still used, but in a different way. There the drinking is a communal social activity, and men will sit around a pot of beer with long straws.

Some of the older traditional drinking customs of Korea are still partially followed today. These mostly centre around paying respect to the elders in a group. If a senior person offers you a drink, then you should stand, if possible, and receive the glass with both hands. Older custom dictates that you should also turn away from them when you are drinking.

Beer is often connected with status in this way in indigenous cultures. For example, in Cameroon, a ceremony called the 'cattle dance' is performed among the Koma people to honour a man's wife as a good mother. This takes the form of a feast, where meat and beer are distributed among others in the community and to the religious leaders. Hosting seven ceremonies will increase the husband's status, and he is then permitted to drink beer from his own pot without having to share it.[20]

Another custom found in many pubs in connection with beer consumption is the 'Yard of Ale' competition. This is simply a challenge to see who can drink the contents of a glass container holding approximately three pints of beer in the shortest amount of time. The yard glass is long and thin, with a large bulb at one end containing the majority of the beer and a flared opening not unlike the horn of a trumpet at the other. There is a trick to drinking the beer that involves rotating the glass as it is held to the mouth. Not doing this leads to pockets of air forming which cause the beer to spill out over the face of the drinker. Historically, many cultures, including the Greeks, the Gauls and most notably Scandinavians during the Viking period, drank beer from horns. These were often indicative of the status of the bearer, ranging from simple hollowed out cow or ox horns to those inlaid with silver and decorated with elaborate carvings.

The yard of ale glass may be seen as descending from the drinking horn in style. Traditionally it was often linked with ceremonies of initiation, either on reaching the age to be able to drink legally or sometimes as part of joining a college fraternity or similar social group. The glass form of the vessel probably originated in the seventeenth century, and was designed for coach drivers who were in a hurry when travelling between destinations and needed to drink quickly. The length of the glass meant that it could be passed to them from a window in the tavern without them having to leave their coach.

Some customs relate to the act of drinking in social groups. The most obvious of these is the method of purchasing drinks for everyone in your group, generally known as buying drinks in 'rounds'. In Australia, this is known as 'shouting' and it is common to hear people both in that country and elsewhere telling others that it is 'my shout' or 'my round'. A clear etiquette has developed around this system, meaning that any individual in the group who does not buy their round will be viewed as being mean with money.

Some methods of choosing who buys drinks are more unusual. Writing on Texas superstitions, George D. Hendricks noted that in one community, anyone in a group who quoted one of the local men who had a reputation for exaggeration or bending the truth in a conversation would be penalised by having to buy the next round of drinks for the group.[21]

Another store was known for a custom that they called 'cutting the book':

> Usually the men came into the store in groups of three or four. As his turn came, each man took a dull-bladed butcher knife, usually kept inside the rather thick book left in prominent display on the counter. With the knife he opened the book. In the presence of the clerk or owner the man checked the page number, paying attention only to the right-hand digit. The one whose bad luck it was to cut the lowest number was obligated to treat the others to a cold drink. If two men had the same low number, then they both cut the book again to break the tie.

Many drinking customs have spread from German- and Austrian-speaking areas. Oktoberfests – traditional gatherings known for having oompah bands and giant mugs of beer, and where many dress in the traditional German lederhosen – are found all over the world, and have gradually become culturally appropriated into other countries. A similar thing has happened with festivals relating to the mythical figure of the Krampus or Perchten (see Chapter One).

BREWING AND BUSINESS

All of these customs and superstitions associated with both the brewing and consumption of beer demonstrate how important a place it had, and continues to have, in society. This is demonstrably the case in other ways too. In many cultures, particularly those where beer is viewed as a food as much as it is a drink, it adds to an individual's daily intake of calories significantly. We have already seen that it may kill the bacteria found in water that has not been treated. Beer also contains a richer mix of vitamins, minerals and protein than unleavened bread.[22]

The production of beer has an obvious economic importance as well as driving these social aspects. As mentioned earlier in the chapter, Egyptian workers were paid in beer. (Interestingly, they were also given onions and we already know that these have become oft cited as a hangover remedy; maybe the Egyptians knew more than we realise!) Many indigenous peoples use beer in the same way as payment to people who work the fields or build houses.

In more developed areas, the brewing process is economically more important as a business. Wherever you find a business, you will then find business rivalry, as one brewery tries to outdo another or make their product seem superior. Some of the less ethical ways of doing this can sometimes leach across into folklore in the form of urban myths.

In 1986 or 1987, a rumour began somewhere in California that the very popular bottled beer Corona was contaminated with urine. The story said that Mexican workers urinated in the bottles and that this gave the beer its bright yellow colour and foamy appearance. The story was traced back to a beer wholesaler, who sold other brands but not Corona, although it is unlikely that they began the rumour themselves. Despite the fact that the practicalities make the story very difficult to believe, if you stop and think about it (Corona produce a massive amount of beer and calculations showed that it would take seventy-five Mexicans 'working together' to contaminate each vat[23]), it soon began to spread across America and caught hold enough that Corona's importer went to the news outlets to deny the rumour.

In recent years, changes have come about in beer consumption that have allowed the market to expand. Craft beers are now increasingly popular, meaning that microbreweries and brewpubs have become more common. Many of these show an interesting interaction between the product (the beer) and place (the location at which the beer is brewed). Beer names are often drawn from popular culture, or from important local folklore to the county or town in which the brewery is sited.

Take, by way of example, Skinners Brewery in Cornwall. Formed as a community brewery in 1997, Steve Skinner's company has now won hundreds of awards for its ales. The first two of these were both named after a local myth (Cornish Knocker) and character (Betty Stogs). The brewery's clever mix of marketing and cultural information summarise both neatly.

Cornish Knocker: the name of the fairy inhabitants of Cornwall's famous tin mines. These mischievous little people appear on our bottles just as they're described in the Cornish tales, with big heads, long arms and wrinkled faces. The mine-dwelling folk would knock to lead miners to the richest seams, so long as the miners left them a little of their pasties when they ate! Nowadays we're pretty sure that they can occasionally be heard in local pubs, knocking on the bottles to guide drinkers to the malty golden ale that bears their name. Just remember to leave a sip behind for them.

Betty Stogs: goes down in Cornish folklore as a convivial character – a woman who would go house-to-house on the Cornish Moors where she lived, entertaining folk. The only problem was she did it at the expense of her poor baby who she left unwashed and uncared for while she went about her wily ways! Eventually the Cornish piskies carried the poor babe off. Lucky Betty Stogs found her child once more in the forest in a bed of moss, cleaned and cared for by the piskies and she vowed to become a more attentive parent. Like our award-winning bitter, Betty Stogs was full-bodied, a little bit fruity and the heart and soul of any party.[24]

Other craft breweries produce beers with similar themes. Wychwood Brewery (whose logo has a witch on a broomstick) has ales including Hobgoblin, King Goblin and Wychcraft. Elgood's brewery has an ale called Black Dog, and in the United States there is even a brewery and mead producer called simply 'Folklore'.

We come full circle to the start of this chapter and the most common folk motif associated with beer and brewing: that of John Barleycorn. There are, as you would expect, many breweries called John Barleycorn, many beers named after him and many pubs with the same name serving the drinks.

So, who was John Barleycorn? Was he a real person? Have all the songs, poems and prose writings drawn on the history of an actual character in their construction? The short answer to this is no. There is no evidence to suggest that John Barleycorn is anything other than a metaphor for the circular growth, harvest and use of barley as a crop discussed earlier. But in folklore terms, there is far more to John Barleycorn than this and so we need to dig a little deeper to find explanations.

The manuscript written by George Bannatyne in 1568 might be the first time that the story is recorded, but we can safely say that it is earlier than that because Bannatyne himself noted that his manuscript was a collection of pre-existing songs and poetry.

The symbolism itself is indicative of the old ways. Even if there is not a direct lineage that can be traced back that far, the story is rich in folk tradition and memory. In writing his classic study of mythology and comparative religion *The Golden Bough*, Sir James Frazer posited that John Barleycorn was evidence of the pagan worship of a vegetation god, who was sacrificed to bring about a fertile harvest. Whilst this leap of faith is too much for some, we can still recognise it culturally in the film *The Wicker Man* and in various folk traditions and festivals. And maybe there is more to the idea than first appears if you look into it.

There are deeper folk meanings hinted at in the poems and subsequent versions recorded in song that point towards the older ways. We can see these right at the beginning of the song:

There were three men come out of the west
Their fortunes for to try
And these three men made a solemn vow
John Barleycorn must die

We know that the number three is culturally significant for many peoples. It is the number of the divine and sacred in many religions. Christianity speaks of the holy trinity. In Chapter One we spoke of the three fates. The number three is also often ascribed magical properties in fairy tales.

The three men coming from the west might lead us to consider a parallel with the biblical story of the wise men who came from the east. The difference in direction is important. In the Celtic mythos, the west stood for the realm of the fairies, or the otherworld. This is obviously reversed in the Christian version to the east to differentiate it, although the mystic element is kept. There are other parallels between John Barleycorn and the Christian idea of birth, life, death and resurrection for Christ. Note that he rises from the dead three days later – another appearance of the magical number.

Moving through the descriptions of growth to the harvest metaphor, we can find more links with ancient practices. When ritual sacrifice was common, the 'Triple Death' was often employed. Drawing again on the sacred ideas of the number three, the victim was killed three times – or at least by three different methods. One of these was usually by hanging. There are a number of bog bodies that have been discovered and which support this idea, such as Lindow Man, Grauballe and Oldcroghan.[25]

We might recognise a similar theme in the story of the crucifixion (crown of thorns, cross and spear in the side) and we also have it in John Barleycorn:

They hired men with scythes so sharp
To cut him off at the knee

They rolled him and tied him by the waist
And served him most barbarously
They hired men with sharp pitchforks
Who pricked him to the heart

Corn dollies were, and often still are, traditionally made as part of the harvest. Also sometimes called corn mothers, they are decorative items woven from some of the ears of corn from the field and retained until the next crop. They are often reminiscent of human shapes, although other designs more akin to works of art may be produced.

Old beliefs said that the spirit of the corn lived amongst the crops. When the harvest was brought in, that would effectively make the spirits homeless. The corn dolly provided the spirits with a home over the winter. When the following spring came around and it was time to plant the new crop, the dolly (and hence the spirits of the corn) would be ploughed into the soil. Because of the human shape of the corn dolly often made, and the fact that it may be tied with a red ribbon, some anthropologists believe that it is representative of this older ritual sacrifice. The ribbon would symbolise the blood and the ploughing into the soil, the burial.

To link back to the brewing trade and beer after all of that, there is some evidence from the analysis of well-preserved bog bodies to suggest that a strong beer-like drink made from barley was given to sacrificial victims before the ceremony to ensure that they were compliant or could not back out. The custom of giving out soul cakes, made from barley, at the Celtic new year celebration of All Hallows Eve (or Hallowe'en) has further parallels. Some cakes would be eaten, whilst some would be kept back and buried under the first soil turned when the field was ploughed in the spring.

And it is to baking, and the milling of the flour needed for it, that we turn for our final subject.

Chapter Four Sources

1. News and Notes No 132, The Oriental Institute of the University of Chicago, Autumn 1991
2. Brooks, J., 'Know Your Beer Gods and Goddesses', www.skinnersbrewery.com/the-myths-and-legends-behind-our-beers Accessed 12.7.19
3. Connelly, A., 'The Science and Magic of Beer', The Guardian, 29.7.11
4. Bloom, J.H., Folk Lore, Old Customs and Superstitions in Shakespeare Land, 1930
5. ibid
6. Cited in On, L.Z., Ancient Beer Recipes Lead to Modern Health Remedies, Newsweek Magazine, 26/9/15
7. Nelson, M., The Barbarian's Beverage: A History of Beer in Ancient Europe, Routledge, 2008
8. Coote Lake, E.F., 'Folk Life and Traditions', Folklore Vol 67, No 4, Dec 1956
9. www.etymonline.com/word/hangover Accessed 22.9.19
10. Paulsen, F.M., 'A Hair of the Dog and Some Other Hangover Cures from Popular Tradition', The Journal of American Folklore, Vol 74, No 292, Apr – Jun 1961
11. Pliny, Natural History, Book xxix, Ch. 5
12. Paulsen, F.M., 'A Hair of the Dog and Some Other Hangover Cures from Popular Tradition', The Journal of American Folklore, Vol 74, No 292, Apr – Jun 1961
13. ibid.
14. Daniels, C.L., & Stevans, C.M., (Eds). 'Encyclopaedia of Superstitions, Folklore and the Occult Sciences of the World, University Press of the Pacific, 2003
15. Bloom, J.H., Folk Lore, Old Customs and Superstitions in Shakespeare Land, 1930
16. Daniels, C.L., & Stevans, C.M. (Eds). 'Encyclopaedia of Superstitions, Folklore and the Occult Sciences of the World, University Press of the Pacific, 2003
17. Smith, O., 'The surprising countries that consume the most beer per capita', Daily Telegraph online, 3 Aug 2018. Accessed 20/9/19
18. Department of the Navy, Naval School, Civil Engineers Corps. Mess Night Manual, August 1986
19. Colicchio, T., The Oxford Companion to Beer, Oxford University Press USA, 2011

20. Arthur, J.W., Beer through the Ages: The Role of Beer in Shaping Our Past and Current Worlds, Anthropology Now, Volume 6, 2014

21 Hendricks, G.D., 'More Texas Superstitions', Western Folklore, Vol 24 No 2, April 1965

22. Arthur, J.W., Beer through the Ages: The Role of Beer in Shaping Our Past and Current Worlds, Anthropology Now, Volume 6, 2014

23. Fine, G.A., 'Mercantile Legends and the World Economy: Dangerous Imports from the Third World', Western Folklore, Vol 48, No 2, April 1989

24. www.skinnersbrewery.com/the-myths-and-legends-behind-our-beers Accessed 26.9.19

25. Scarre, C., (Ed). The Human Past, Thames and Hudson, 2009

MILLING AND BAKING

Miller, miller, take up the grain
Pour it out like sand
Miller, miller, open the rill
To turn the wheel and work the mill
Grind the grain to flour and fill
The sacks below.

Shine the sun
And rain the rain
Fall the shivery snow
Hail and wind and sleet again
As the year will go.

Baker, baker, the flour is here
Soft and fine and bland
Baker, baker, get out of bed
Put that silly old hat on your head
Bake me a loaf of golden bread
And then I'll go

Song lyrics instead of poetry open this chapter, and I confess that it is a personal indulgence. These verses come from 'The Miller's Song', written and performed by Sandra Kerr and John Faulkner. They are taken from the classic 1970s children's TV programme *Bagpuss*. In this show, a girl named Emily owns a shop populated with toys that are brought to life by magic. The shop does not sell anything, but rather houses lost items for people to come and collect. The objects are brought in broken and the characters in the shop repair them and learn about them. In the episode 'The Mouse Mill', a model of a watermill is brought in. Once repaired, the mice that fixed it try to fool the other characters into believing that the raw ingredients that go into the mill will produce chocolate biscuits automatically.

Bagpuss is a much-loved programme, known for its intersection with folk tales and practices from around the world. The song in this case describes the process of baking from field to end product; it is calendrical in the same way as the John Barleycorn story from the previous chapter.

MYTHOLOGY AND HISTORY

Baking is arguably one of the oldest methods of cooking, and the products connected with it are generally associated with grain. Going right back to prehistoric times, humans had worked out the transition from heating or roasting seeds, to combining them with water to produce a mixture that could be put onto a hot stone and cooked. Later, enclosed ovens led to the production of loaves rather than flatbreads.

The switch from unleavened to leavened bread probably came about when dough left around before being baked started to react with yeasts in the mixture. Over a period of time and observation, this process became understood and controllable, leading to the baking methods that we know today. This reaction in the dough is a type of fermentation, which links the baking process very much to that of brewing. Over time, craftspeople working with the two items learned that barley was the best grain for beer production, and yeast the best for baking.

Although baking was well established in early Egypt, it was much later that it reached Rome, and Pliny the Elder recorded in the second century that there were no bakeries there. Professional bakers only began to be established when the women who lived in wealthier families were desirous of avoiding the tedious job of making bread.

To start at the beginning of the process: you cannot bake bread without flour, and for that you need a method of being able to grind the corn down. The first examples of milling were done by hand using quern-stones. A quern was made up of two round stones, one fixed and one mobile. The bottom stone was stationary and was the actual quern. On top of this was another stone, known as the handstone, which was rotated by means of a shaft slotted into a hole on the outer edge. Grain was introduced through a central hole called a hopper (the precursor of that part of actual mill machinery) and would then be ground between the two stones.

The British Museum has evidence to suggest that the quern was first used as a means of milling by hand as early as the Neolithic period. Archaeological investigation has found their use all over the world. Grain querns have been uncovered in China; in Mesopotamia they were used for grinding a whole range of edible and non-edible materials. American Indians are also recorded as having used querns for games, as well as for their normal practical use. In terms of folklore symbolism, the movement of the quern was sometimes seen as representative of the cosmos and in turn, from that, the constant turning of the seasons. Myths from Finland and the Scandinavian countries speak of the World Mill, constantly turning out cosmological good or bad fortune.

In the *Kalevala* – the epic poem of Finnish folklore – it is told that Ilmarinen, a smith, forged a device called the Sampo for the Mistress of Pohjola. This mystical item had three sides and three functions (you may recall that we spoke in earlier chapters about some of the magical properties of the number three and their later religious significance). One side of the Sampo produced salt, one produced coins and the third ground corn. The finished Sampo was taken to the Mount of Copper and protected with nine locks. It put three roots out: one into the earth, one into a mountain and one into the sea. This has parallels with the mythology of the Scandinavian World Tree Yggdrasil. Another tale from Denmark connected with this method of milling offers explanation as to why seawater contains so much salt.[1] King Frodi buys two very large and strong female slaves called Fenia and Menia from the King of Sweden. In many versions of the myth they are referred to as 'giantesses'. In Denmark there are two millstones that have magical properties. They will produce whatever the person grinding them desires them to. However, they are so big that nobody has been able to move them, and so the magic has never been used. King Frodi has Fenia and Menia brought to the mill, named Grótti, where the stones are located. There, he orders them to grind him gold and prosperity and will not let them rest until a song is sung. The two women, however, sing a song called Gróttasavngy, meaning 'The Mill Song', and grind an army to fight against Frodi.

That night, a sea king called Mysing comes to the mill. He kills King Frodi and takes Fenia and Menia and the mill away with him on a ship. He orders the women to grind salt and they continue doing so until midnight, when they ask whether he has enough. He tells them to carry on grinding and after not much more time has passed, the ship becomes too heavy and it sinks into the sea, along with the millstones, the salt and Fenia and Menia. This, says the tale, is what caused the sea to have salt water. Furthermore, there was always a whirlpool in the sea at that place thereafter, being brought about by the water entering the eye of the stone. A version of the folktake from the Orkneys puts this site at a whirlpool called 'The Swelchie' in the Pentland Firth.[2]

What is interesting about this particular myth is the variety of places to which it has travelled and re-rooted itself. The book *Superstitions of Sailors* notes that a similar story may be found at the *Ille et Vilaine* in Brittany:

Once upon a time there lived a sorcerer who had invented a mill that could grind anything that the sorcerer commanded it to. The mill would only stop when the inventor pronounced a certain formula. One day a mariner heard of this wonderful mill and stole it. When he reached the open seas, the mariner commanded the mill to grind a quantity of salt which he required for the codfish he was out to catch. Soon the vessel was full of salt, but, alas, the mariner, ignorant of the magic formula, had no power to stop the mill in its work. One and on the mill continued to grind large quantities of salt so that the vessel and the mill sank to the bottom of the sea under the heavy weight. The mill is still continuing to grind the salt, and hence the salty taste of the sea water.[3]

The *Historia Brittonum*, or *History of the Britons*, was written around 828AD, probably by Nennius, and was a key source for Geoffrey of Monmouth, as well as being the first text to contain King Arthur. The Welsh monk describes thirteen wonders of the country in the book. One of these is named as the 'Wonderous Mauchline Quern', which he said would grind continually every day, except for on the Sabbath. It is thought by some that the location Auchenbrain may derive from this story, as it means 'field of the quern'.[4] Others have placed the magical quern much further away, in Kent.

The story of Fenia and Menia is illustrative of the fact that milling using a hand quern was normally undertaken by two women, who would often sing traditional songs whilst they worked. This is mentioned in one of the Biblical parables concerning the Day of Judgement, wherein Jesus says that 'two women shall be grinding at the mill; the one shall be taken, and the other left'.[5] One of the oldest surviving pieces of writing, credited to Antipater of Syria in 85BC, referring to a mill also refers to women operating it:

Traditional windmill in Kent. Photo: Tracey Norman

Ye Maids who toiled so faithfully at the Mill
Now cease from work and from your toils be still.
For what your hands performed so long and true,
Ceres has charged the water nymphs to do.
They come, the limpid sisters to her call
And on the wheel with dashing fury fall
Impel the wheel and with a whirling sound
They make the massy millstone turn around
And bring the floury heaps luxurious to the ground.[6]

PRIMITIVE MILLING

The quern is the primitive forerunner of the more practical mechanical mills that were to follow on, first in terms of watermills, which began to emerge around the second or third century AD, and then windmills, following on and becoming recognisable around the eleventh or twelfth century.

Primitive mills did not have a miller working in them. The arrangement at this time was that people with grain would simply go to the mill, and the mill would grind it into flour. It is at this point that we find that mills begin to become associated with fairies: it is thought that the creatures do the work overnight in the same way as sprites help in houses or farms. Lady Wilde describes such a tale in her 1887 book on Irish legends and folklore:

> One day a farmer's son [named Phadrig] was minding cattle in the field when something rushed past him like the wind; but he was not frightened, for he knew it was the Phouka on his way to the old mill by the moat where the fairies met every night. So he called out, 'Phouka, Phouka! Show me what you are like, and I'll give you my big coat to keep you warm.' Then a young bull came to him lashing his tail like mad; but Phadrig threw the coat over him, and in a moment he was as quiet as a lamb, and told the boy to come to the mill that night when the moon was up, and he would have good luck.
>
> So Phadrig went, but saw nothing except sacks of corn all lying about on the ground, for the men had fallen asleep, and

no work was done. Then he lay down also and slept, for he was very tired: and when he woke up early in the morning there was all the meal ground, though certainly the men had not done it, for they still slept. And this happened for three nights, after which Phadrig determined to keep awake and watch.

Now there was an old chest in the mill, and he crept into this to hide, and just looked through the keyhole to see what would happen. And exactly at midnight six little fellows came in, each carrying a sack of corn upon his back; and after them came an old man in tattered rags of clothes, and he bade them turn the mill, and they turned and turned till all was ground. [7]

It is a natural progression to lead on from stories concerning fairies to other kinds of supernatural encounters in the mill building. For a long time, the mill would be the only building in a village with machinery that moved seemingly of its own power (but in actuality due to the forces of nature in water or wind form). This added a further supernatural quality to the mill, and so it should be no surprise that stories of haunted mills are not uncommon in folklore.

MECHANICAL MILLING

In the early development of these mechanical mills, farmers would bring their grain to be milled rather than buying flour directly. It is with the advent of these buildings that the folklore surrounding millers really begins to establish itself.

Mills became tenanted buildings belonging to the local lord or landowner, and therefore a rental economy began to form, with a paid miller being put in place to control the use of the mill and to collect a toll from the customers wishing to use the facilities. Hand milling using quernstones was obviously seen as a direct threat to this as they prevented the miller from taking the toll (or tax) from the people bringing their grain to be milled. Legal grants were given in England and Scotland for millers to be able to go out and break any quernstones that they found, and which were therefore preventing them from exacting the toll for their landlords. This practice actually went on for a much longer time than you would imagine. In Ireland, use of the quern carried on for many more years and a correspondent to the *Dublin Penny Journal* in 1836 notes:

> I think it was in the year 1794, Mr Wm. Acheson of Roscagh, who was then proprietor of Kesh Mill, in the county of Fermanagh, gave orders to his miller, Stephen Belford, to break all the querns he could find. The only pair I then remember in the neighbourhood, belonged to one Patrick McManus ... and when any neighbour would borrow the quern, it was as carefully concealed from Stephen Belford, as an illicit still would now be hidden from an exciseman.[8]

THE MILLER AS A VILLAIN

The early image of the miller in folklore is that of a crook or a villain. There are a number of reasons for this that we should consider. Working in an important role at the centre of village life, in a trade that requires special skills, will always set a person apart. We have seen this already in the first chapter with regards to the weaver, and also in the chapter on blacksmiths. It is the same with the miller. Here, the mill has already developed something of a reputation because of the magical nature of the machinery. There is a supernatural association with the building, which therefore carries on to the figure of the miller. And, as is the case with the other craftspeople, in folklore this usually means that the devil is involved somewhere along the line.

With the miller however, the devil is not so much responsible for giving special skills to the miller (after all, it is the mill itself that is doing all the work in this case). Here, rather, the miller usually does a deal with the devil that contributes somehow either to his wealth, or to his villainy. This is the case in a folk tale called 'The Girl Without Hands'. It is not one of the more generally known tales, mostly because of its . slightly dark nature, which means it doesn't tend to be told to children at bedtime!

In the story, a miller who is living in poverty meets the devil disguised as a man. The devil promises great wealth to the miller, and in return he asks for what is behind the miller's shed. The miller agrees to this, because he believes that it is the apple tree behind the shed that the man is wanting. It is, however, the miller's daughter who is behind the shed, and on whom the devil has his eye. The miller's daughter is an innocent girl, however, and every time the devil comes for her he can find no reason to take her away with him. After this has gone on for some time, the devil becomes very angry and gives the miller an ultimatum: either he must cut off his daughter's hands, or the devil will take the miller instead. Exhibiting the more unpleasant side for which the character of the miller is going to become renowned, he does indeed cut off the hands of his daughter.

Many years later the daughter, who has been made free, marries the king and they have a son together. But the devil once again tries to interfere in order to get his prize, and orchestrates events to try and make everyone believe that the king wants his wife dead. Fortunately, the king's mother refuses to kill the girl and her son and sets them free so that nobody can find them. Seven years later, the king tracks them down. The girl's hands have miraculously grown back and, as in all good fairy tales, they live happily ever after.

There are, of course, moral messages around the tale concerning greed. Folk tales often carry morality aspects within them. Another tale involving a miller and the devil concentrates on this important aspect of levying taxes, which did so much to make the character of the miller unpopular and mean in folkloric terms. The tale tells that a miller had the option of taxing people heavily or lightly depending on who they were, as he ground corn for farmers at all social levels. The devil stood at the miller's shoulder and conversed with the miller on how he should toll the farmers.

The first farmer to come to the mill was very rich and had fifty wagons of corn. The miller asked the devil how he should toll the rich farmer.

'He probably got rich being hard on the poor,' said the devil. 'Toll him heavy.'

The next farmer to visit had ten wagons of corn. He was neither particularly rich, nor particularly poor.

'He is neither rich nor poor,' the miller noted.

The devil explained to the miller that the man would get along just fine. He was healthy and happy, and had no real problems in life. 'Toll him heavy,' said the devil.

The last man to arrive at the mill carried just a bushel of corn on his back. The farmer showed the devil how poor the man was, and then once again asked how he should toll him.

The devil replied, 'He is poor. Keep him poor. Toll him heavy.'

So the miller did indeed toll the man heavy.

Taxes were, and always will be unpopular, and the tolling of farmers by a greedy miller is forever enshrined in a traditional folk song generally known as 'The Miller's Sons'. The song is listed as number 138 in the Roud Folk Song Index, and one of two versions held at the Bodleian Library, probably from the eighteenth century, tells the story in the following verses:

There was a Miller who had three Sons,
And knowing his life was almost run,
He call'd them all, and ask'd their will,
If that to them he left his mill.

He called first for his eldest son,
Saying, My life is almost run,
If I to you this mill do make,
What toll do you intend to take?

Father, said he, my name is Jack,
Out of a bushel I take a peck,
From every bushel that I grind,
That I may a good living find.

Thou art a fool, the old man said,
Thou hast not well learn'd thy trade;
This mill to thee I ne'er will give,
For by such toll no man can live.

He call'd for his middlemost son,
Saying, My life is almost run,
If I to thee the mill do make,
What toll do you intend to take?

Father, says he, my name is Ralph,
Out of a bushel I take it half,
From every bushel that I grind,
So that I may a good living find.

Thou art a fool the old man said,
Thou hast not learned well thy trade;
This mill to you I ne'er can give,
For by such toll no man live.

He called for his youngest son,
Saying, My life is almost run,
If I to you this mill do make,
What toll do you intend to take?

Father, said he, I'm your only boy,
For taking toll is all my Joy.
Before I will a good living lack,
I'll take it all, and forsware the sack.

Thou art my boy, the old man said,
For thou hast well learned thy trade;
This mill to thee I'll give, he cry'd,
And then he clos'd up his eyes, and dy'd.

The reference to the sack in the ninth verse of this song is very obscure, but is possibly explained in a newspaper article from June 1957, which was noted in the *Western Folklore* journal.[9] The article, which was telling stories of the founding fathers, noted that one of them named Bill Soden established a mill on the Cottonwood River. One day a friend played a trick on Bill, after which he ran down the street waving an empty grain sack. As the miller chased the man it prompted some children watching to cry, 'Look how Bill Soden has robbed a man of all his grain and is now chasing him for the sack.'

An old proverb compares the nature of the miller with some of the other skilled craftspeople: 'Put a miller, a tailor and a weaver into a bag and shake them; the first that comes out will be a thief'. This is not the only proverb that refers to the miller as a villain. Another says that something may be 'safe as a thief in a mill', implying that a thief could safely hide in a mill because the miller would be the bigger thief of the two.

It appears that the miller has these villainous aspects to him in other countries in the world also. In Bulgaria, for example, it was believed that flour needed to be fumigated using incense after the wheat was brought back from being ground at the mill. This was especially important if the miller was a dark man, because this meant that the flour may have been bewitched. As well as doing this, some of the flour should be dropped on the floor as you entered the house. This would prevent any demon who may have been in the flour from entering also. It was also thought to be unlucky in Bulgaria to sell a sack of flour unless a loaf of bread had been baked using it first. This seems to be some sort of superstitious quality control system.

In Persia, it was said that when you brought home the first flour that had been ground in a season, then a piece of iron should be put into the bag. This resonates with the magical qualities of iron that we looked at in the chapter on blacksmiths. Here, placing the iron in the flour would stop devils from coming and stealing it away. Iron was not the only thing that Persians believed was good to put into your flour. They also believed that if you put a twin almond (two kernels in one shell) into your barrel of flour, then this would act as a preservative, and would make the flour last five months longer than it usually did.

SUPERNATURAL QUALITIES

Sometimes, the supernatural nature of the mill comes to the forefront of the stories. A strange tale comes from Quakertown in Pennsylvania, around the year 1846. The water-powered mill there was run by Peter Riedman, who was frequently milling until eleven at night because so many people wanted their grain ground. One night, being exhausted from so much work, the miller told his son that the following morning he should go to the mill and open the sluice gates ready for the day. He did as he was asked, but ran back home scared saying that he could not get into the mill because it had turned around during the night. Sure enough, when Peter went to the building himself, he saw that it had indeed turned around. Being a superstitious man, Peter believed that this was probably the work of witches.

The phenomenon was certainly a great talking point in the town. People travelled from miles around to come and see the mill. A local carpenter said that he believed that the mill could be turned back again if enough people helped out, using a series of tools and levers. Many of the local farmers, whose names are recorded in reference to this event, agreed with the idea. Eventually, some 160 people gathered together to assist and, after much work, they did indeed manage to turn the mill around. Interestingly, it was said that after this was done the mill actually ran better than it did before the incident occurred.

This is a curious story and it is difficult to say how much basis in fact or otherwise it has. In the account of it that is written up in the journal *Midwest Folklore*[10] many names are recorded and it is noted that some of them at least do appear in the historical record. The article also states that there were no unusual weather events on the night in question, and that the mystery will probably never be solved, suggesting that it is not seen purely as a folk tale or legend.

Witches are mentioned in another way connected to a mill, known as the 'crooked mill' in Monastir, Macedonia. The name derives from the fact that the millstone turns in the opposite direction to that usually found in a mill. Here, it is said that you can find out whether you have been bewitched (although it may be the case that a miller at one time landed upon a great money-making opportunity!). Between the Thursday of Passion week and Ascension day, people who want to discover this fact head to the mill every Thursday. They take a stork's egg that they crack, and then use the shell to take some water from the mill stream, with which they wash their faces. Next, they go into the mill and sit on the millstone, which the miller turns for them. It is said that the stone will turn easily if the person is not bewitched, whereas if they are it will not turn unless the person pays a sum of money to the miller. The more they pay, the faster the stone will turn.

Another odd custom here is said to help a girl who has not been able to find a husband. She should go to the mill with a new water jug, which she should use to take some water from the stream, making sure that she does not do so against the current. This water should be thrown over herself, after which she should sit in the sun to dry. Once dry, the girl needs to go to the roof of the mill and turn over some of the tiles before running away, without looking back. She will then be successful in her quest.

Stones that were said to be imbued with sacred power may be found at a mill at Killin in Perthshire. These are from a mill said to have been founded in the ninth century by St Fillan, originally for the weaving industry but then later as a grain or meal mill. It is said that St Fillan used to stand on the bridge next to the mill and preach sermons to the local people, who stood on the pebbles below to listen. Over time, these pebbles became sacred to those who stood there, and they used to be used to massage people who had medical conditions or diseases. After the death of the saint, the pebbles were collected and continued to be used for massage. Over time, many were lost, but the remaining eight were placed into a niche in the mill wall by one of the Lords. Another version of the story regarding these stones says that people had a superstitious belief that the stones would treat sick cattle if you immersed them in water and then gave that water to the animals to drink. This is not too dissimilar to some of the folklore concerning the blacksmith's water, which we looked at in Chapter Three. Women were also said to use these stones to treat diseases of the chest.

An unusual superstition held at one time by the Pennsylvania Dutch was that the mill could be used as both a preventative measure and a cure for whooping cough. This strange remedy centred around the belief that a child who is afflicted should be placed in the hopper of the mill with the grain. They should be left in the hopper until all of the grain has been ground out. Although documented in the nineteenth century, interviews conducted within living memory have attested to people doing this.[11] An elderly couple in 1962, for example, said that they took their daughter to their local mill in 1930, when she was suffering from whooping cough, and placed her in the hopper. Their neighbours had advised them to do this. Several mill owners spoken to around the same time also said that it had been quite common for parents to bring their children to the mill for this purpose.

There is a breed of moth known as the Miller, getting its name from the fact that it leaves a white residue on the hand if you pick it up. This matches up with the trope of the miller being covered in white flour, from where the commonly heard nickname 'Dusty Miller' emerges. There is also an evergreen shrub with silvery-white leaves with the same name.

Miller Moth. Photo: Ben Sale www.flickr.com/ photos/33398884@N03

In the county of Dorset, an old children's game performed a sort of ritualised killing of the Miller moth, which could be seen as a response to the mistrust of the miller as a community figure who had enough importance to be able to levy taxes. Children would catch the moth, question it and the toll that it had taken and then condemn it to death with a rhyme:

Millery, millery, dousty poll,
How many zacks hast thee a-stole?
Vow'r-an'-twenty, an' a peck –
Hang the miller up by's neck[12]

The symbol of flour residue can also be found in a story that tells of how a tree stump at Westhall in Suffolk came to be known as 'The Miller's Stump'.

On a very dark and misty night in 1763, a miller called John Turnbull was carrying a sack of flour home. He had not been able to grind the grain at his own mill of Fulwell because the sails were broken, but the flour was needed urgently. Stopping for a rest on the way, he leant the sack of flour against the trunk of a large tree and sat down.

When he was ready to continue on his way again, the miller picked up the sack and put it on his back. But, in doing so, he caught the sack on a branch of the tree, making a small hole in it. As he walked, a trail of flour from the hole in the sack formed on the ground behind him.

The local squire, in his horse and carriage, was driving down the same road and spotted the trail of flour. Because of the misty conditions, the white flour served as a useful guide to the road and so the squire urged his horses on to follow the trail, not realising that the miller was not very far ahead of him. The carriage hit the miller, who was killed, but the squire had not realised and so did not stop. John Turnbull's body was not found until the next day.

On hearing of the accident the following day, the squire visited the location and saw that the large elm tree where the miller had rested grew in such a way that it stopped people from seeing down the road properly. He therefore had the tree cut down, and ever since the stump has been called the Miller's Stump. The widow of the miller cursed the stump, saying that it would never again grow tall. And of course, when the night is dark, it is still said that the ghost of the miller will continue his journey to deliver the sack of flour.

Legends attached to a deceased miller can also be found at an interesting gravesite called 'The Miller's Tomb' at Highdown Hill in Sussex. The miller in question here is one John Oliver, a wealthy man who died in 1793. He has always been viewed in diverse ways by locals in the area. Some believe that he was a little mad, others that he was very religious, and more that he was traditionally villainous.

Oliver is buried in a table-top-style tomb, surrounded by iron railings on the top of the hill. The mill and his house also used to stand here, but both are now gone, and so just the tomb remains as a strangely out of place structure. Prior to vandalism in the 1980s, the tomb was inscribed on every surface with religious verses and Biblical extracts.

Although probably a somewhat honest and pious man in life (albeit prosperous), it is only twenty years or so after his death that less savoury aspects of the miller's life begin to be asserted. A nineteenth century history of that area of Sussex contains the following text about John Oliver:

> ... he certainly was a very singular character. He amused himself a good deal in the construction of machinery – two pieces of which afforded the spectator some merriment. The one represented a mill and a miller, so constructed that every time the shafts were moved by the wind a sack opened and a shovel was in the act of raising the flour for the purpose of filling it! The other represented a Customs House officer with upraised sword pursuing a smuggler, and an old woman at the heels of the officer, violently banging him with a broom! At that period this part of the coast was much frequented by smugglers, and some have suggested from the above circumstances that the honest miller had a secret predilection for these gentry.[13]

By calling the miller honest, the author is probably suggesting that this was his general outward appearance to people, but that underneath he was working to help the smugglers, and that perhaps this is another reason why he was wealthy. The folklorist Jacqueline Simpson notes that there has been speculation that the mechanical figures, which might remind you of the wind-driven garden ornaments that were popular in the 1970s, were used as a method of signalling the smugglers. She also states that other Sussex millers have been said to have been 'in league with' smugglers, but that the accusers do not normally have any evidence to back up this rumour.[14]

Two particular legends are associated with this tomb. One is that the ghost of the miller may be raised by walking a number of times around the tomb, after which he will appear and chase you. Sometimes this is stated as twelve, but more usually as seven (one of the magical numbers in folklore and myth). This is not unique, and is a trope that may be found at a number of other tombs in Britain. Usually it comes about as some kind of initiation, or children's dare game.

The second story, which is almost certainly apocryphal, says that the miller was buried upside down so that when the Day of Judgement came and the world was turned upside down, he would be facing the right way when the dead were called from their graves for reckoning. This idea is, again, not unique and may be found elsewhere in parody of the Christian practice of burying bodies with the head to the west of a grave and the feet to the east, because Christ would come from the east on the Last Day and so the dead could rise to face him. In these stories, which are somewhat mocking, the deceased person is usually a local character who is disliked because of their wealth or position in the community.

THE HONEST MILLER

When mills began producing flour from their own grain to sell rather than milling the grain brought by farmers, then the symbol of the miller as an untrustworthy rogue or villain gradually began to wane. Without the levy of tolls in the same way, there was not such an unpleasant aspect upon which to latch. The change is reflected in the less-unsavoury nature of such things as folk songs from later periods. For example, the 'Miller of Dee', often known as 'The Jolly Miller' (Roud folk song #503), comes from the city of Chester, on the banks of the River Dee. It was written for the play *Love in a Village* by Isaac Bickerstaffe in 1762 and the original lyrics were:

There dwelt a miller, hale and bold, beside the river Dee;
He danced and sang from morn till night, no lark so blithe as he;
And this the burden of his song forever used to be: -
'I care for nobody, no not I, if nobody cares for me.

'I live by my mill, God bless her! she's kindred, child, and wife;
I would not change my station for any other in life;
No lawyer, surgeon, or doctor e'er had a groat from me;
I care for nobody, no not I, if nobody cares for me.'

When spring begins his merry career, oh, how his heart grows gay;
No summer's drought alarms his fear, nor winter's cold decay;
No foresight mars the miller's joy, who's wont to sing and say,
'Let others toil from year to year, I live from day to day.'

Thus, like the miller, bold and free, let us rejoice and sing;
The days of youth are made for glee, and time is on the wing;
This song shall pass from me to thee, along the jovial ring;
Let heart and voice and all agree to say, 'Long live the king.'

In this song, there is no mention of the previous dislike for the miller as a character. In fact, sometimes the miller proves to be the cleverer person, such as in a folk tale with a few variants known as 'The Miller at the Professor's Examination' or similar titles.

The story centres around a famous professor from overseas, who comes to England to examine all of the college students in the country. His last place of examination was at Cambridge, and the students there were all very nervous about his visit because he was so eminent.

As he journeyed to the university, they tried to trick him to prevent his visit. One group of students dressed as labourers and, when the coachman stopped to ask them the distance to the town, they answered in Latin. Another group further down the road did the same in Greek. But the plan backfired. The professor decided that, if the working classes were so educated as to speak Latin and Greek, then the students must be excellent in Cambridge. So, on arriving at the university, he told everyone there that because of this fact he would examine them using signs rather than language.

One student in particular had been studying very hard, and was expected to be the top of everyone tested. He was despondent because this announcement made the playing field level. Even the most lazy student would have the same chance of guessing the professor's signs as him.

On the day of the examination, the student took himself off and sat by the river next to the mill. The miller, seeing the student there and knowing him from talking before, asked what the problem was. The student explained about the visiting professor and his plan to use signs in the examination. The miller told the student that he had nothing to worry about.

'Did you never hear that a clown may sometimes teach a scholar wisdom?' asked the man. He told the student to swap clothes with him and that he would take the exam in his place. If he passed, the student would get the credit. But if he failed, the miller would reveal who he was so that the student would not suffer the failure. The student pointed out that everyone knew that he only had one eye, but the miller said that he would wear a black patch, and then nobody would be the wiser.

At the university, the professor had tested every student but not one had passed. And so came the turn of the miller, disguised as the student.

First, the professor put his hand into his coat pocket and took out an apple. The miller put hand into his own pocket, pulled out a crust of bread and held it out. Next, the professor put the apple back and pointed at the miller with one finger. The miller pointed at him with two fingers in return, to which the teacher responded with three fingers. The miller finally held out his fist, after which he was told that he had passed the test.

Leaving the room, the miller ran back to his mill and told the student of his success. They changed clothes and the student ran back to the university to collect his prize. In the lecture room, the professor explained the meaning of the signs in his test.

'First,' he said, 'I held out an apple, signifying thereby the fall of mankind through Adam's sin, and he very properly held up a piece of bread, which signified that by Christ, the bread of life, mankind was regenerated. Then I held out one finger, which meant that there is one God in the Trinity; he held out two fingers, signifying that there are two; I held out three fingers, meaning that there are three; and he held out his clenched fist, which was as much as to say that the three are one.'

The student received his prize, but he could not understand how the miller had been able to unravel the test when none of the students could. Returning again to the mill, the student explained to the miller everything that the professor had said, to which the miller replied with how he had interpreted the test:

'When I went in the professor looked mighty fierce, and he put his hand in his pocket, and fumbled about for some time, and at last he pulled out an apple, and he held it out as though he would throw it at me. Then I put my hand in my pocket, and could find nothing but an old crust of bread, and so I held it out in the same way, meaning that if he threw the apple at me I would throw the crust at him. Then he looked still more fiercely, and held out his one finger, as much as to say he would poke my one eye out, and I held out two fingers, meaning that if he poked out my one eye I would poke out his two – and then he held out three of his fingers, as though he would scratch my face, and I clenched my fist and shook it at him, meaning that if he did I would knock him down. And then he said I deserved the prize."[5]

The piece of bread that figures in the folk tale of the miller and the professor leads us on to consider the folklore that may be associated with the end product, which comes from the flour produced in the mills. Although, of course, there are many things that can be baked using flour, bread is one of the simplest and culturally the most significant. It is such a major staple food in the diets of those all around the world that it is not surprising that much myth and superstitious lore and ritual have formed around bread.

BREAD

It should be noted at this stage that it is very difficult in many ways to separate the folklore of bread from that of the raw grain. Bread may be seen as symbolic of metamorphosis (from seed to grain to loaf) and of cooperation (farmer to miller to baker). It therefore has a cyclic pattern in the same way as does the growing and harvest of the grain, as we saw in the last chapter with the John Barleycorn legend. However, as far as possible, the focus will primarily be on bread itself as a product of grain in this examination.

Bread has a long-standing religious significance, aside from the vegetation gods associated with the fields and harvest. In the Jewish faith, it is much evident during Passover, or the Feast of Unleavened Bread. During this period of observance, no other leavened products are permitted in a Jewish home, in remembrance of the Jews who had to flee from Egypt so quickly that they could not wait for their bread to rise.

Many people would naturally think of the communion when thinking of bread, and in the New Testament, the story with Jesus and the disciples is also one of remembrance:

> Then Jesus sat with his disciples at the table of the feast and said, Behold the lesson of the hour: ... Then Jesus took a loaf of bread that had been broken not and said, This loaf is symbol of my body, and the bread is symbol of the bread of life; And as I break this loaf, so shall my flesh be broken as a pattern for the sons of men; for men must freely give their bodies up in willing sacrifice for other men. And as you eat this bread,

so shall you eat the bread of life, and never die. And then he gave to each a piece of bread to eat. From henceforth shall this bread be called Remembrance bread ...

Here we have both the element of remembrance within the story itself, but also a remembrance of much older pre-Christian celebrations where rituals had the body of a god sacrificed and distributed between participants.

The Mexican holiday of *Dia de los Muertos* – the Day of the Dead – is an important time during which families come together to honour their dead relatives and pray for them in their spiritual journey. This is a day of celebration for those of Mexican heritage rather than sadness, as the dead awaken to celebrate with them. Much like other folkloric celebrations that used to take place in a single locality, such as those involving the Krampus in Austria at Christmas, the Day of the Dead is increasingly seen elsewhere as it is culturally appropriated (or, as some would argue, mis-appropriated) in other countries. Particularly common as a symbol is the colourful sugar paste skull known as an *alfeñique*, which is placed on a decorated altar as an offering. Bread is also used as an offering during this celebration. During the weeks leading up to *Dia de los Muertos*, a circular bread called *Pan del Muerto* (Bread of the Dead) is baked and then placed as an offering with the sugar skulls and candy pumpkin. All of the food will be eaten after the ceremonies are concluded, but some believe that the food itself has no nutritional content by then, as the spirits of the deceased will already have consumed its spiritual essence.

In Bulgaria there is much religious significance placed upon bread in the lead up to Christmas. This is because the Day of St Ignatius of Antioch, known as Ignazhden and celebrated on 20 December, is when 'the young year starts'.[16] At this time we find traditions very similar to those of first-footing, regarding the first person to visit the house in a new year. In Bulgaria, if the first person to enter a home is good and wealthy, then they will bring well-being to the household for the coming year. On Ignazhden therefore, visitors will not be allowed in if they were not invited.

Bulgarians traditionally believe that the Virgin Mary's labour began on Ignazhden, and for this reason it marks the beginning of the holidays. Many ritual breads are baked during this time. On the morning of Ignazhden, the women prepare dough for baking. One custom is to take a hemp comb and remove part of the dough with it. They would then knead this dough but it would not be baked, set aside instead to dry. It would later be used as a traditional cure for someone who was ill, given with honey. The same dough might also be used prior to baking for ritual house protection. Using two fingers, some of the dough would be rubbed in a cross sign on one of the wooden beams of the house to dispel evil. After all this was done, the dough would be used to make one loaf for every member of the house, plus an extra loaf for Christmas Eve.

Different traditional loaves would also be baked for the table on Christmas Eve. One would be decorated with a cross and was dedicated to God. Another would be decorated with images of haystacks and would be dedicated to agriculture. Again, we see more parallels here with the harvest celebrations. In some places it is believed that bread baked on Christmas Eve will never become mouldy.

Using bread as protection, as with the dough in the example above, is not uncommon. In Jerusalem during Ramadan, bread was hung in the house as a protective amulet. Often, this would be made in the shape of a hand with the fingers outstretched. Both the hand and the bread were seen as protective in this case. Putting a cross onto the bread is generally seen as protective for the people who will eat it. Certainly, in some parts of France it was generally considered to be unlucky to eat bread before it had been marked with the sign of the cross.

One old belief in Palestinian areas, which survived through to modern times in memory at least, is that the tree of 'knowledge of good and evil' from the story of the Garden of Eden was a wheat tree.[17] After being expelled from Eden, Adam and Eve had to subsist on the product of cursed soil, being mostly thorns and thistles. God eventually had pity on Adam and sent the archangel Gabriel to him with wheat grains in silk handkerchiefs. Because the soil was so poor, the wheat would only grow to the height that we now see in our fields, rather than the much more bounteous crops that were originally to be found in the Garden of Eden. This act of God marked him out as the provider of our most basic food requirement and to this day in the Lord's Prayer we still say, 'give us this day our daily bread'.

Due to all of this ritual connection and religious importance, it was said that it was profane to waste or abuse bread in any form. For this reason, crumbs of bread were (and often still are) collected up and thrown out as food for the birds or other small animals. In this way, people were seen to be providing for God's creatures. Conversely, anyone throwing bread into the fire, or destroying it by other means, would soon be found to be wanting in life.[18] To throw bread crumbs into a fire was to feed the devil, it was believed. Punishments were worse in Turkey – there, to step on a piece of bread meant that you would be sent to the third hell, where you would spend eternity being gored by an ox.

Although it is good to make bread on Christmas Eve, there are other times of the year where it is not so favourable to bake. You should not make bread on All Saints' Night (Hallowe'en) because the veil between the worlds is said to be thin at this time, and if you do bake then ghosts will eat the bread. Superstition also said that it is impossible to make bread on Good Friday, because the water needed to mix it will turn into Christ's blood at this remembrance of the crucifixion. In the way that folklore is so often contradictory, however, other superstitions are more favourable with regards to Good Friday. Some people say that it is on that date, rather than Christmas Eve, that bread does not go mouldy. And in many countries, a loaf that is baked on Good Friday morning and kept until the following year is said to be a treatment for stomach problems. The stale bread should be grated into a glass of water, which should then be drunk.

Depending on what you do with bread, folklore ascribes either good or bad luck to you. It is said to be good luck to carry a crust in your pocket, but it is bad luck to take the last piece of bread from a plate – unless you are a bachelor. If that is the case, then it was thought that your chances of marrying a rich woman would be significantly increased. In France, it was said to be bad luck to put a loaf of bread on the table upside down. Elsewhere, there was also a superstitious belief that if you laid the bread down on the table on its side then something bad would follow.

Norwegian superstition has bad luck being the outcome if a piece of bread is thrown onto the floor. This may be connected to the older ideas of wasting bread, which were connected to religious aspects. Sweden and Germany also have this belief. It is also said to be unlucky to step on a piece of bread, and further, that if you drop a piece of bread on the floor and then pick it up and eat it you will be poor thereafter. We can see that there is a very strong connection with wasted bread on the floor. In Estonia, if a piece of bread was dropped, it would be picked up and kissed as a mark of respect.

All of these ideas stem from the idea that bread is a gift from God, as we have already discussed, and that to degrade it in any way will lead to significant punishment. A Middle Eastern folk tale illustrates this:

> A peasant woman went to the tābûn (a type of clay oven) to bake bread. She took her baby along with her. After baking a few loaves the baby's bowels moved. Having no rags whatsoever the mother cleansed him with a hot loaf of bread. As a punishment for dishonouring this heavenly gift the child was turned into a monkey with its gluteal region red. Since that time all monkeys have this characteristic.[19]

It is interesting to note that in some versions of this story, it is added that God sent the angel Gabriel to the woman with seven silk handkerchiefs for her to use to cleanse the child, but that instead she decided to keep them for herself and use the bread. This adds an element of greed, rendering it in part a story with a moral, as so much folklore is. But also, it obviously has links to the Garden of Eden story, as mentioned earlier, wherein Gabriel appeared with wheat seeds that were also enclosed in silk handkerchiefs.

There are several pieces of lore that treat bread as a divinatory food in some way. For example, if it crumbles while being cut, then an argument of some kind is likely to follow. This is preferable to the consequences of a belief in some areas of Appalachia, USA, which say that if the first cut into a loaf of bread slices through a hole, then it is a portent that someone will die. In a similar vein, the English county of Yorkshire it was believed that if your bread did not rise during baking then it was a sign that there was a corpse somewhere nearby waiting to be found. In the Highlands of Scotland, it was thought that singing while you were baking was the action that caused the bread to fail to rise, whereas a superstition over the border in England stated that it would not rise if there was a corpse in the house.

Macedonian folklore beliefs say that it is a sign of death if the baker forgets to put salt into the bread, and also if water escapes from the dough while it is being mixed. Another broader superstition says that if there is a crack in a loaf of bread when it has been newly baked, then there will soon be a death in the family. Also connected with death is one old belief from Suffolk: that if it was expected that someone would soon pass away, then it was sensible to have a large quantity of bread prepared so that evil spirits would be too busy eating it to worry about the soul of the deceased, which could then leave in peace. This is perhaps similar to the idea of playing on a vampire's compulsion to count by scattering breadcrumbs on the ground, so that they are kept occupied and therefore are not a threat.

These ideas seem to be cross-cultural. At Ramadan, many believers in rural areas and in the towns and cities would throw a handful of seven (that important number again) different kinds of cereals over the threshold of their houses. The cereals used comprised wheat, barley, millet, emmer (a type of hulled wheat), lentils, beans and vetch (a member of the legume family whose seeds must be processed before being eaten, like the kidney bean). They did this because of a belief

Statue of Egyptian woman making bread. Photo: Rama, CC-by-sa-2.0-fr

that the djinn, banished from the mortal realm during the holy period, would return at the end of Ramadan. In this case, there is not a connection with counting as with the vampire. Rather, the threat to the djinn was twofold. Firstly, they would slip on the seeds and hurt themselves in the fall. And secondly, by treading on the wheat (from which bread, the gift of life, is derived) they would be dishonouring the Almighty, as in the other examples above.

It is not too surprising that there is a lot of folklore joining bread and death, when you consider that archaeological evidence has many cultural examples of bread in a funerary setting. Both grain and bread were put into the sarcophagus by Egyptians to provide nourishment for the deceased on their journey to the afterlife; Hindus would also give bread to the deceased; and in Japan, people place bread on graves as an offering. Through to the end of the nineteenth century it was common for bread to be given out as a dole at funerals. This often took place at the back of the church, and some buildings had a 'bread bench' for this purpose.

There is an unusual connection to be found between bread and hair – unusual because it is different for dough and the baked bread itself. It was said that if a woman who had dough on her hand stroked a young man's face, then he would subsequently never be able to grow a beard. But if someone ate a large amount of bread, then that was said to mean that they would develop a chest full of hair.

Folklore hasn't only formed around the baking of the bread and the way that it is used – it even extends so far as to details such as how you cut it. And it is important that bread is broken or cut in some way, as one superstition told that if a loaf left the table uncut, then the people at the table would leave hungry. If you cut one slice too many at the table however, then you should expect a stranger to turn up and eat it. Some believed that if you wanted to become prosperous or successful in life, then you should cut the top side of the loaf first, making sure that the cut is even and not rough. In fact, to cut bread in an uneven fashion generally was said to be a sign that the person cutting had told lies during the day at some point.

Christian peasants in the Near East would never cut bread at all, but would break it apart. This custom was in respect to Christ, who never cut bread with a knife. A peasant invited to have breakfast by a pastor whom he was visiting one day, told the pastor off for cutting the bread with a knife, asking him why he believed himself to be better than Jesus, who always broke his bread.

Superstition related to bread has even gone so far as to enter situations where bread is not even present – in a similar fashion to it being unlucky to walk underneath a ladder, in America couples or groups would say 'bread and butter' as a blessing when they are split up by an obstacle such as a lamppost whilst walking. Saying this phrase was supposed to prevent ill fortune or an argument from following. The root of this probably comes from the fact that you cannot split bread and butter apart once a piece of bread has been buttered.[20]

As well as being a gift from God, bread has also traditionally been a gift between people. At the New Year customs already mentioned, bread would be one of the items brought into the house by a visitor after midnight on the new day. Here the symbolism was that the house would never know hunger. In the same way, it is also given on some occasions as a housewarming gift. In Estonia, bread was traditionally given along with salt. In Macedonia, it is traditional to give a new bride a loaf of bread and a bottle of wine, with which she goes to her new husband's house. The symbolism is similar to that of the New Year traditions – ensuring that the house will always have what it needs. Brides there were also taken straight to the kitchen before they went to any other room in the house, carrying a loaf of bread under each arm. This would ensure that the household was wealthy. Once in the kitchen, the wife would search the ashes of the fire in order to find a small wheat cake, which would have been hidden there by her husband's mother. This would bring her the skills to ensure that she was a good bread maker in the future.

Bread has even been presented as a gift to the royal family. This happened in an unusual form for the Queen Mother on a visit to Edinburgh in 1956, as reported in *The Times* newspaper:

> When Queen Elizabeth the Queen Mother and her sister Lady Elphinstone were accorded the freedom of the historic burgh of Musselburgh, Midlothian ... they were the first women to be so honoured. An amusing feature of the ceremony was the presentation to the new burgesses of Musselburgh's traditional 'bap', the origin of which is obscure but is believed to be a link with the days when the local fishermen received gifts of fancy bread on leaving harbour. The baps were embossed with the burgh arms.

History does not seem to recall what the Queen Mother did with the bap!

Finally, some pieces of bread superstition are as unusual as the gift of a bap. In Persia, it is said that you show disrespect to God if you eat bread with your head uncovered. This is not unusual at all. What is more curious is the belief there that if a man puts bread on his head, he will die by famine. There is also folklore in Russia that says putting bread on your head will result in a bad harvest. Neither of these countries are particularly known for the carrying of goods on the head, so it is not clear why people would be considering putting the bread on their head in the first place.

And a warning should you ever visit Portugal – if you have bread that has just been taken from an oven, you should never put it to your nose to smell it. If you do, after you have died, worms will eat only your nose.

Chapter Five Sources

1. Thorpe, B., 'Northern Mythology: comprising the principal popular traditions and superstitions of Scandinavia, North Germany, and the Netherlands', Edward Lumley 1852

2. www.orkneyjar.com/folklore/pentland.htm Accessed 25.9.19

3. Rappoport, A.S., 'Superstitions of Sailors', Stanley Paul & Co, 1928

4. Watson, William J., 'The History of the Celtic Place-Names of Scotland', William Blackwood and Sons Ltd, 1926

5. Matthew 24:41 or Luke 17:35

6. Scott, W.M., 'A Hundred Years A-Milling: Commemorating an Ulster mill centenary', Carter Publications, 1951

7. Wilde, Lady, 'Ancient legends, mystic charms and superstitions of Ireland', Ticknor and Co, 1887

8. Anonymous, 'Ancient Irish Hand-Mill, or Quern', The Dublin Penny Journal, Vol 4, No 193 (Mar. 12, 1836)

9. Burress Jr, L.A., 'The Miller and his Three Sons', Western Folklore, Vol 21 No 3 (Jul 1962)

10. Bond, L., 'From the Archives: Unsolved Mystery of the Old Brown Mill', Midwest Folklore, Vol 11 No 3, Indiana Issue, III (Autumn 1961)

11. Smith, E.L., and Stewart, J., 'The Mill as a Preventive and Cure of Whooping Cough', The Journal of American Folklore, Vol 77 No 303 (Jan – Mar 1964)

12. Udal, J.S., 'Dorsetshire Children's Games etc', The Folk-Lore Journal, Vol 7

13. Evans, J., 'Picture of Worthing', 2 vols, London: Chambers, 1864

14. Simpson, J., 'The Miller's Tomb: Facts, Gossip and Legend', Folklore, Vol 116, No 2 (Aug 2005)

15. Wright, T., 'The Miller at the Professor's Examination', The Folk-Lore Record, Vol 2 (1879)

16. Bezovska, A., 'Ritual bread-making from Day of St Igantius until Christmas', online at https://bnr.bg/en/post/101060260/ritual-bread-making-from-day-of-st-ignatius-until-christmas Accessed 29.9.19

17. Canaan, T., 'Superstition and Folklore about Bread', Bulletin of the American Schools of Oriental Research, No 167, (October 1962)

18. Gregor, W., 'Bread', The Folk-Lore Journal, Vol 7, No 3 (1889)

19. Canaan, T., 'Superstition and Folklore about Bread', Bulletin of the American Schools of Oriental Research, No 167, (October 1962)

20 Beckwith, M.W., 'Signs and Superstitions Collected from American College Girls', The Journal of American Folklore (Jan – Mar 1923)